COLD-SM[OKING]
& SALT-CURING
MEAT,
FISH,
& GAME

Other Cookbooks by A. D. Livingston

Sausage
Cast-Iron Cooking
The Whole Grain Cookbook
Venison Cookbook
The Curmudgeon's Book of Skillet Cooking
Jerky
Complete Fish & Game Cookbook
Freshwater Fish Cookbook

C1

COLD-SMOKING
& SALT-CURING
MEAT,
FISH,
& GAME

A. D. LIVINGSTON

LYONS PRESS
Guilford, Connecticut
An imprint of Globe Pequot Press

To buy books in quantity for corporate use
or incentives, call **(800) 962–0973**
or e-mail **premiums@GlobePequot.com**.

Lyons Press is an imprint of Globe Pequot Press.

Project editor: Gregory Hyman
Layout: Kevin Mak
Text design: Elizabeth Kingsbury

Library of Congress Cataloging-in-Publication Data is available on file.

ISBN 978-1-59921-982-0

Printed in the United States of America

10 9 8 7 6 5 4 3 2 1

Contents

Acknowledgments

The author would like to acknowledge material quoted from *A Man's Taste,* reprinted by permission of The Junior League of Memphis, Inc.; from Mary Randolph's *The Virginia House-wife,* first published in 1860; and from Cy Littlebee's *Guide to Cooking Fish & Game,* reprinted by permission of the Conservation Commission of the State of Missouri. For different perspectives on smoked and cured meats, the author thanks *The South American Cook Book* by Cora, Rose, and Bob Brown; *The Country Kitchen,* a British cookbook by Jocasta Innes; *Bull Cook and Authentic Historical Recipes and Practices* by George Leonard Herter and Berthe E. Herter; and *Butchering, Processing and Preservation of Meat* by Frank G. Ashbrook. These and other acknowledgments are also made in the text, as appropriate.

Preface

I have written a dozen introductions for this book. Some long, some short. Some direct, some oblique. All of these earlier drafts seemed too serious for the text, and maybe this one is, too. Curing and smoking meats, fish, and game at home ought to be fun, and the results ought to be culinary delights, or at least be a welcome change from supermarket fare. In order to cure and cold-smoke meat safely, however, salt is required. Lots of salt.

Unfortunately, *salt* has become a bad word in the culinary and health-food trade. The trend these days is for writers and TV reporters and marketing experts to pussyfoot around the issue or to capitalize on it by treating salt in the negative. As a result, a lot of modern people suffer from what I call salphobia. I wrestle a round or two with this problem in Chapter 1 because I feel obliged to do so to the best of my ability. At this point I want to say two things: In the short term, skimping on the salt used for home-cured and smoked meats and fish can be very dangerous to your health; indeed, an unsalted turkey put into an electric smoker during windy or cold weather, along with a pan of water to keep the moisture up, can be a veritable salmonella factory. In the long term, cutting out salt-cured and home-preserved meats has brought us to rely more and more on supermarket fare. Read the newspapers. People die of food poisoning. Chicken has become a toxic substance. We are told to cook everything until well done, even prime T-bone beefsteak. We are told to wash our hands thoroughly after handling meat. We are told to spray the countertop with Lysol. The situation is so bad that we hear more and more about zapping supermarket foods with radiation. Safely, we are told. Our "cured" hams are already embalmed with water and chemicals. Safely, we are told.

But all this is heavy stuff. I don't want to clutter my mind with it, and I resent having to burden this book with it. Consequently, I have decided to front the issue in Chapter 1 and get it out of the way. Then I'll get on with some hopefully enjoyable information about curing and smoking meats at home, along with definitions and nuts-and-bolts how-to text on such topics as cold-smoking, hot-smoking, salt-curing, sugar-curing, and so on.

But what if I am wrong about salt? Well, in that case, I'll have to recall a spirited discussion that I once had with a fun-loving doctor who made some money and took it to Alaska. The gist of the conversation was that society simply can't afford to keep people alive forever, a matter that our politicians will have to face sooner or later. The good doctor's solution was that everybody should be issued a book of tickets. When all the tickets are gone, he said, that's it. Well, it seems to me an equitable way to run things, giving a break to people who die accidentally in their youth, or in war, and punishing those of us who tend to burn the candle at both ends. In any case, when the new deal goes into effect, I'll surely spend two tickets, if necessary, on a Virginia ham.

—*A. D. Livingston*

PART ONE

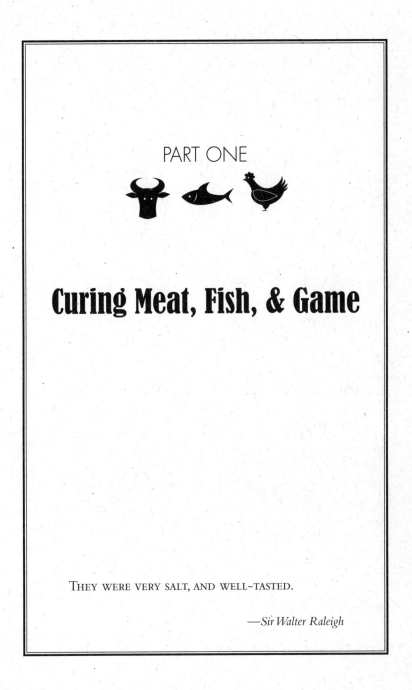

Curing Meat, Fish, & Game

1

THE SALT IMPERATIVE

When our daughter comes home for a visit, she watches me like a hawk while I am in the kitchen. The girl has a problem. On the one hand, she is starved for good home cooking and a return to the old ways; on the other hand, she lives in mortal fear of common table salt. It's a miserable dilemma, in that you simply can't have good cooking without a little salt. It's a gustatory necessity.

The trouble, I think, is that the girl fell in with the wrong crowd. Health-food freaks. Yuppies. Vegetarians. New Agers. Meditators. I tell you, some of our young people are too gaunt and thin and sickly. The worst of them eat no meat and run or exercise a lot—a highly dangerous combination simply because most vegetables contain very little salt and because sweating depletes the body's supply of the mineral. If these people feel weak and tired and crave dill pickles, they may very well need some salt instead of vitamins and herbs and zucchini and meditation. Why? Why else would the U.S. Navy have advised me to take a large salt tablet daily when I worked in the hot engine room of the USS *Donner*, LSD-20? Why else would the U.S. Army have issued salt tablets to soldiers in Desert Shield and Desert Storm? Because the human body must have salt to function properly, that's why. It's biochemistry.

Is it possible that the American press—magazine articles, advertisements, popular health books, and TV shows—has greatly misled the public? I think so, and I have support from a modern

British book called *Food in History*, written by Reny Tannerhill. In addition to requiring salt to function, we humans may also be conditioned psychologically to salt. As Tannerhill says, "Modern health advisors who would ban all salt from the diet are, in effect, recommending rejection of a substance that humanity has been genetically and socially programmed, over a period of tens of thousands of years, to desire and need." I can't repeat the whole argument here, but, from a historical viewpoint, salt has been man's best friend in the mineral world. Without salt we might not have been able to domesticate grazing animals, which must have salt and which know this instinctively and travel long distances to salt licks. One theory has it that animals were first domesticated near the cave entrance, where human urine concentrated salt.

Without salt, and before mechanical refrigeration was invented, man could not have preserved meats, the lack of which would have greatly restricted travel and sea voyages. Also, salt was by far the most important item in early trade. Many of the desert trails still lead to sources of salt. The modern Via Salaria of Italy was first made by Roman carts carrying salt. The word *salary* came from the Roman word for salt allowance, or ration. Indeed, salt has been used as money from Abyssinia to Tibet, and it has been an important source of tax revenue even in modern times, as Mohandas Gandhi well knew.

In any case, I personally want and need some salt in my food, and I prefer to dine with people who feel the same way. Of course, I don't want to give advice in this matter, except to say that anyone who quits eating salt because of media medicine, Richard Simmons, or new wives' tales really ought to consult a doctor of medicine. Even then, a second opinion may be in order. If the human body is functioning properly, any excess

salt that may have been ingested through normal eating practices will be eliminated quite naturally in perspiration and urine.
Naturally, there may be limits to how much salt one can safely
eat, but the taste buds and common sense are usually safe guides
in this matter. Of course, if your doctor tells you not to add any
salt whatsoever to your daily food, then you are clearly reading
the wrong book.

Meats and fish have been cured for many centuries with
the aid of salt, and sometimes with both salt and smoke. From
time to time other ingredients were added. For a while saltpeter
(a term that usually means potassium nitrite) became popular,
but it has been more or less replaced by sodium nitrite, sodium
nitrate, and other compounds such as ascorbic acid. Some of
these substances are used primarily to preserve the color of the
meat, and the meat trade has even resorted to various dyestuffs.

If we look back at the folk cures of America, we see that
there's more than one way to cure meats. The *Foxfire* books,
which became something of a hit some years ago, contained an
interesting collection of meat cures. The material for these books
was gathered in the Appalachian Mountains of North Georgia
by students from a rural school at Rabun Gap. Unfortunately,
most of the cures are set forth far too sketchily to be of much
practical value. Even so, I find the material to be very interesting, and I hope that the summary below helps establish the fact
that curing meat does not have to be an exact science and can
be accomplished without pumping good meat full of water and
chemicals and dyestuffs and Liquid Smoke.

At hog-killing time in *Foxfire* country—usually the first cold
snap in the fall of the year—the meat to be cured was cooled
down and salted as soon as possible. Some people would simply
cover each piece of meat with salt and let it cure for a suitable

length of time; usually, but not always, the salted meat was arranged on a wooden shelf or packed into a wooden box or barrel for curing. Taylor Crockett used exactly 8 pounds of salt per 100 pounds of meat. He mixed the salt with 1 quart of molasses, 2 ounces of black pepper, and 2 ounces of red pepper; he smeared this mixture onto the meat and let it cure for 6 to 8 weeks, depending on the weather.

Valley John Carpenter used 5 pounds of plain salt for a 200-pound hog, and Ron Reid used 10 pounds of salt per 100 pounds of meat. Most of the *Foxfire* people left the salted meat on wooden shelves in the smokehouse to cure, although a few packed it into boxes or barrels. Lake Stiles took the meat down into his cellar and put it on the dirt floor, allowing the earth to "draw the animal taint out of the meat, keep it cool, and prevent souring or spoiling."

If the meat was not to be smoked, the *Foxfire* people usually left it in the curing mode for the winter, during which time they could simply take out what they needed and cook it. In the spring the unused meat was washed and treated again. At this time some people covered the meat with a mixture of black pepper and borax to keep the skipper fly larvae out. Bill Lamb put a mixture of borax and black pepper on the washed meat, then smoked it. Others used brown sugar and pepper, put the meat in a cloth bag, and hung it in the smokehouse. Lake Stiles washed the meat and then buried it in a box of hickory ashes, whereas his grandmother used cornmeal instead of ashes. Another fellow—who allowed no one to spit into the fireplace—sifted the ashes, put them onto the dirt floor of the smokehouse, and buried the meat in them. Lizzie Carpenter put some freshly shelled white corn into the bottom of a wooden box, then added middling meat, another layer of corn, and so on until the box was full. She claimed that the corn draws the salt out and gives the meat a good flavor.

Of course, some of the *Foxfire* people smoked their meat after first putting it through a salt cure. This was usually accomplished in a walk-in smokehouse, which was considered to be an essential structure at most homesteads. The smokehouse usually had a dirt floor. More often than not the fire was built on the dirt, or perhaps in a small trench, but sometimes an iron wash pot was used as a firebox. Although the *Foxfire* book indicated that the fire was made of small green chips of hickory or oak, pieces of hickory bark, or even corncobs, my guess—based on experience with old-timey home smoking—is that long-burning green limbs or logs of hickory were also used.

The early settlers on the barrier islands of North Carolina, called the Outer Banks, also made very good use of both pork and fish, and both were cured in the fall of the year. At first, hogs were allowed to roam free on the islands, gathering acorns and whatever they could find to eat—and the houses were fenced in to keep the hogs out of the yards. Of course, each settler marked his hogs with a coded series of notches in the ear. Hog-killing time was something of a social event, usually held at first frost or first freeze, at which time the neighbors gathered to help each other. Almost all of the hog was used, from snout to tail. The hams, shoulders, and bacon were salted and smoked, and the trimmings were made into sausage and smoked. The methods of curing pork are similar to those set forth in *Foxfire* and won't be repeated here.

Salted fish was a great staple of the Outer Banks until recent times, and the tradition lives on there simply because the taste for salt fish can't be satisfied by other means. A book called *Coastal Carolina Cooking* sets forth a few favorite methods:

Bill and Eloise Pigott of Gloucester corn their spot (a small member of the croaker family) in brine. They gut the fish but leave on the scales. The fish are placed in a light brine for several days in a barrel, then they are washed and placed in a heavier brine. The spot are taken out as needed—Eloise advises one to avoid using a metal spatula to remove fish from the brine, as metal will taint the whole batch. Before being cooked, the fish are freshened by soaking them for some time in fresh water.

Mitchell Morris of Smyrna uses a dry cure for spot, layering them with salt in a barrel. The fish stay in the salt until 24 hours before cooking, at which time they are placed in fresh water to soak. Mitchell also salts roe from mullet, which are ready in October and are best, he says, on a full moon. He salts the sacs and presses them between unpainted wooden boards in a sunny place for several days. He eats the roe fried, baked, scrambled, or plain, saying, "Old folks put roe in their pocket and eat it like candy."

I too am fond of mullet roe—or any good roe of small grain—salted in a similar manner (see Batarekh, page 27).

Although the techniques are easy—requiring only salt, time, and a relatively low temperature—the authors of *Coastal Carolina Cooking,* who worked for the Sea Grant program, a federally funded project, felt obligated to say, in a sidebar, "Even though coastal residents still use these methods of preserving seafood, seafood specialists discourage their use unless you are experienced in the techniques, as improperly preserved fish could spoil. Experts recommend you freeze your catch instead."

The same sort of disclaimer was inserted into the *Foxfire* books. The chapter on curing hog, for example, starts off, condescendingly I think, with the following paragraph: "Meat was cured by the mountain families in several ways. Professional

8

butchers today would probably shudder at the apparently hap-hazard measurements they used, but they often seemed to work." Well, I suspect that some of the old practitioners from the mountain families would shudder at the modern methods of pumping chemicals and water into good hams. Some of the old-timers might even marvel that the new methods "often seem to work," in spite of constant reports in the media these days about food poisoning and salmonella.

I could sketch in a hundred folk cures from other parts of the country, all with variations in method and ingredients. But the old cures all have three things in common: lots of salt, plenty of time, and relatively cool weather. And that's all you need to cure fish and meats at home. Smoke can be added to the process, as discussed in Part Two, but it is more for flavor and is usually not a necessary part of the curing process. Other ingredients, such as sugar and spices and garlic, can be added to alter the flavor of the end product, and chemicals such as sodium nitrite or ascorbic acid are often added to preserve the color. But salt is the only ingredient that is really needed for curing the meat. Omit it at your peril.

2

CURED FISH

As pointed out in Chapter One, salt fish has been very important to the world's expanding population. During the Middle Ages and later, a great commerce developed. The nations of Europe rose and fell by their control of the herring fishery of the North Sea. Later, salt cod from Newfoundland and New England became very important.

In addition to preservation, another reason for salting fish was to reduce its weight. Fresh fish are mostly water, and the salting process draws out a good part of it. In fact, salting and drying reduces the overall weight of fish by as much as 80 percent. The volume is also reduced. This fact was important to anyone transporting fish by mule train, sailboat, or dogsled.

Even today, salting the catch will often help with storage and transport problems, since the salt cure requires no refrigeration or ice. But culinary considerations are perhaps more important for modern man. As the recipes in this chapter show, properly prepared salt fish have a unique flavor and have become part of the regional cuisine of some areas. Even the texture of a familiar fish can be altered by a salt cure. Soft fish such as crappie or spot, for example, can be made much firmer. Also, bony fish such as suckers, pickerel, and shad can be salted to advantage; the salt softens the bones, making these fish easy to eat.

There are some rather fierce regional salt-fish favorites, such as salt mullet along the Gulf of Mexico, salt cod along the northeastern seaboard, and salt herring in the North Sea countries, but the techniques for salting the various kinds of fish are pretty much the same from one area to another. There are, of course, a number of recipes for salting fish at home or in camp, and champions of this or that exacting technique will no doubt stand firm, regardless of what I say. But the plain truth is that if you leave the fish in contact with enough salt for a long enough period of time, it's hard to go wrong from a curing viewpoint.

How much salt? Lots of it. It's best to buy inexpensive salt by the bag—25-, 50-, or even 100-pound sizes. Almost any salt will do, even that used to de-ice roads and sidewalks.

Salting Techniques

Anyone who has a burning culinary or historical interest in salt fish should read *Fish Saving: A History of Fish Processing from Ancient to Modern Times,* by C. L. Cutting. In this tome are many salting techniques, mostly of commercial application. Since the advent of mechanical refrigeration and more speedy transportation, the commercial trend has been toward a lighter cure. Also, the cure is often combined with light smoking in modern times.

In spite of a wealth of historical and regional formulas, I feel that the modern practitioner, and certainly the novice, can get by nicely with the simple methods below.

A Basic Salt Cure

Catch lots of good fish, preferably in the 1- to 2-pound range. Low-fat, white-fleshed fish like large crappie, walleye, and black bass are good when salted, and so are the fatty fish like mullet and mackerel. Dressing the fish for salting is easy. Without scaling or skinning, merely cut a slab of fish off each side of the backbone, lengthwise, as when making a fillet cut. Of course, you will cut close to the bone, getting all the meat you can. That's it. No gutting or beheading. Leave the skin and scales on. Wash the fish in a solution of 1 cup of salt to a gallon of water. Then drain the fish.

Find a wooden, plastic, or other nonmetallic container of suitable size and shape. Wooden chests are traditional, but large plastic or Styrofoam ice chests are easy to use at home, in camp, or even in a boat. Put a layer of salt in the bottom. Also place some salt in a separate container such as a plastic dish or tray. Lay each fish on the salt in the tray, turn it, and pick it up by the tail, bringing out as much salt as sticks to it. (Also put some salt into the body cavity if the fish has been dressed in the round.) Place a layer of salted fish, skin side down, atop the salt on the bottom of the chest. Then add another layer of salt, a layer of fish, and so on. The last layer of fish should be placed skin side up, and should be well covered with salt. Lots of salt. Cover the container and put it in a cool place for a week. (A basement is ideal. I have also put them under the house.) A brine will develop as the salt draws the water out of the flesh. In short, the fish will shrink in size and become firmer.

After a week, remove the fish and discard the brine. Make a new brine by boiling some water and adding salt to it until the solution will float a raw egg. While the new brine is still hot, add a few peppercorns if you want them. Cool the brine. Put the fish back into the container and pour the cooled brine over them. A weight of some sort, such as a glass platter or a block of wood, should be put on top so that the fish will not float and be exposed to the air. Never use a metallic weight. Cover the container with a cloth and leave it in a cool, dark place for 2 weeks. The fish can be stored for longer periods in the brine, or they can be removed, washed, packaged, labeled, and frozen until you are ready to use them.

Old Dominion Pickled Herrings

Here's an interesting cure that provides a method of taking the fish out of the water and putting them into the brine without having to waste time gutting or cleaning them, which makes it a good technique for curing fish that are taken in large numbers. The method is quoted in full from Mary Randolph's *The Virginia Housewife,* first published in 1860. The words "brine left of your winter stock for beef" refer, of course, to the brine left over from salt-curing beef; if this sounds just too frugal, remember that salt was harder to come by those days. Of course, the modern practitioner might prefer to use fresh brine. (If not, see the Old Virginia Corned Beef cure on page 78.)

The best method for preserving herrings, and which may be followed with ease, for a small family, is to take the brine left of your winter stock for beef, to the fishing place, and when the seine is hauled, to pick out the largest herrings, and throw them alive into the brine; let them remain twenty-four hours, take them out and lay them on sloping planks, that the brine may drain off; have a tight barrel, put some coarse alum salt at the bottom, then put in a layer of herrings—take care not to bruise them; sprinkle over it alum salt and some saltpeter, then fish, salt, and saltpeter, till the barrel is full; keep a board

over it. They should not make brine enough to cover them in
a few weeks, you must add some, for they will be rusty if not
kept under brine. The proper time to salt them is when they are
quite fat: the scales will adhere closely to a lean herring, but
will be loose on a fat one—the former is not fit to be eaten.
Do not be sparing of salt when you put them up. When they
are to be used, take a few out of brine, soak them an hour or
two, scale them nicely, pull off the gills, and the only entrail
they have will come with them; wash them clean and hang
them up to dry. When to be broiled, take half a sheet of
white paper, rub it over with butter, put the herring in, double
the edges securely, and broil without burning it. The brine
the herrings drink before they die has a wonderful effect in
preserving their juices: when one or two years old, they are
equal to anchovies.

A. D.'s Wooden-Plank Cure

I once salted down a batch of small fish—6- and 7-inch golden
shiners—on a wooden plank. The fish were scaled, beheaded,
washed, and drained. A layer of salt was sprinkled onto the plank.
The fish were dredged in salt, then put down in a single layer
without touching. Salt was piled on top, and the board was tilted
a little in the deep sink in our laundry room so that the moisture
could run off.

After a few days, some of these were freshened and fried. I
found them to be quite tasty, and, as I hoped, they could be eaten
bones and all. For the sake of gastronomic research, I washed the
salt from the remaining shiners, put them back onto the board,
and placed them under an air-conditioning vent for drying.

These were kept for several months without signs of rotting, but finally my wife claimed that they were starting to smell and she wanted them out of her laundry room. Since they were bone-dry by now (and really didn't smell), I put them into a plastic container and hid them under the counter in the kitchen. She ran across them one day about a year later and said, in gist, that she wanted them out of her house. The fact that the fish were still edible seemed to make no difference to her. Some women are just hard to live with. In any case, dried fish are covered later in this chapter.

After experimenting with the shiners, I used the same wooden-plank method to salt down some mullet fillets and again to salt a few sucker fillets. The method worked nicely, and I recommend it for salting a few small fish or fillets.

RECIPES FOR SALT FISH

Ironically, some of the best recipes for salt fish were developed in West Africa, the West Indies, and eastern South America—far from the codfish banks of Newfoundland and the North Sea. Why? During the world's colonial period, salt fish was shipped by the ton to these parts from Europe, Newfoundland, and New England. The people developed recipes for cooking the hard salt fish, often with the aid of native ingredients, and developed a taste for them. Even today, in an age of mechanical refrigeration, salt fish are quite popular in many parts of the world. I like them, too, and I remember that my father was fond of eating fried salt fish for breakfast.

Before cooking, most of the salt is removed by soaking the fish overnight in several changes of cool water. This is called freshening. Still, anyone who objects to the salty taste should avoid the fried fish recipes and try those that contain other ingredients—especially potatoes, which absorb some of the salt. In any case, the recipes that follow should contain something for everybody. When serving, remember that salt fish is firm and rich and quite filling, so large portions aren't required.

Salt Fish Mauritius

One of my favorite salt-fish recipes comes from halfway around the world. In the Indian Ocean, 500 miles east of Madagascar, sits the small island of Mauritius. Over the years its people developed a superb blend of flavors with salt fish and other ingredients.

    1 pound salt fish
    ¼ cup peanut oil
    2 cups fresh cherry tomatoes, halved
    1 medium onion, chopped
    6 green onions with tops, chopped
    2 cloves garlic, minced
    1 tablespoon chopped fresh parsley
    1 teaspoon finely grated fresh ginger
    cooked rice

Wash the salt fish, remove the skin, and flake the meat from the bones in large chunks. Put the fish pieces into a glass or non-metallic container, cover with water, and soak overnight in the refrigerator. Change the water a time or two if it is convenient to do so.

When you are ready to cook, drain the fish and pat dry. Heat the peanut oil in a large frying pan and sauté the fish chunks for

5 or 6 minutes. Then add the onion, garlic, parsley, chopped green onions, and halved cherry tomatoes to the frying pan with the fish. Heat and stir until the onion is soft. Add the ginger and stir. Cover and simmer for 15 to 20 minutes. Spoon the fish over fluffy rice. This dish is quite rich and will serve 4 people of ordinary appetite.

A. D.'s Fried Salt Suckers

This technique works for most bony fish, but suckers are my favorite in spite of the bad press they have received over the years. These are salted by the method given in the first part of this chapter. Before salting, however, the suckers are filleted. Each fillet is "gashed" with a sharp knife on a diagonal, placing the cuts about ½ inch apart. Do not cut all the way through the fish but do cut through the layer of Y-bones. With a little practice, you can feel the bones as you cut.

    salt suckers or other bony fish
    peanut oil
    white cornmeal, finely stone-ground
    buttermilk
    pepper

Soak the salt fish in cool water all day or overnight, changing several times. Scale the fish and soak them in buttermilk for 4 hours. Drain the fish, sprinkle each piece lightly with pepper, and shake them in a bag with cornmeal. Heat at least ½ inch of peanut oil in a frying pan. (Or rig for deep-frying if you prefer.) The oil should be very hot, but not smoking. Fry the fish for a few minutes, until they are nicely brown on each side. Drain each piece well on brown grocery bags or other absorbent paper. Eat while hot.

Salt Fish Halifax

Here's a recipe from the Nova Scotia Department of Fisheries. The official version calls for salt cod, but any good salt fish will do.

1 pound salt cod
2 cups mashed potatoes
¼ cup finely chopped onion
¼ cup finely chopped fresh parsley
¼ cup grated cheddar cheese
1 teaspoon black pepper
cooking oil
dry bread crumbs

Freshen the fish by soaking it overnight in cold water. Change the water once or twice if convenient. Simmer the fish in a little water for 10 or 15 minutes, until it flakes easily when tested with a fork. Drain the fish and flake the flesh off the bones. Mix the fish flakes, mashed potatoes, chopped onion, parsley, cheese, and pepper. Form the mixture into patties. Roll each patty in bread crumbs. Heat about ¼ inch of oil in a skillet and fry the cakes for 2 or 3 minutes on each side, or until golden brown. There's enough here to serve 3 or 4 people.

Variation: For fish balls, add an egg to the mixture and shape it into small balls instead of patties. Deep-fry in very hot oil. Drain well on absorbent paper before serving.

Scandinavian Salt Fish

The Scandinavians developed a way of preparing salt fish with sour cream, and several other peoples, especially in the Middle East, cooked salt fish in cream or milk. Here's my version:

1 pound salt fish, boned and skinless
1 tablespoon butter
¼ chopped onion
¼ cup sour cream
1 tablespoon chopped fresh parsley
1 tablespoon fresh lemon juice
⅛ teaspoon white pepper

Soak the salt fish in fresh water overnight or longer, changing the water several times. Flake or chop the fish and drain it well. Melt the butter in a frying pan and sauté the fish flakes for 6 or 7 minutes on high heat. Drain on brown paper. In a serving bowl, mix fish, sour cream, onion, parsley, lemon juice, and pepper. Serve cold on crackers. Feeds 3 or 4.

Solomon Gundy

Here's an old recipe that calls for salt herring. Other salt fish can also be used.

2 salt herring
2 large onions, sliced
2 or more cups hot vinegar
2 tablespoons brown sugar
½ teaspoon black pepper

Wash the salt herring. If they are whole, fillet and discard the back-bone. Cut the meat into chunks and soak overnight in cool water, changing a time or two. Drain and rinse the pieces. Layer the fish in a deep bowl, alternating with a layer of onions. Cover with water, then pour the water into a measuring bowl and note the amount. Discard the water and place an equal amount of vinegar into a saucepan, then stir in the brown sugar and black pepper. Bring the vinegar to a light boil, then pour it over the fish. Cover the container, then cool it in the refrigerator for several hours. Serve cold.

Salt Fish Breakfast

My father was fond of eating salt fish for breakfast, and I too like them after 9 o'clock, along with fresh sliced tomatoes. In rural Florida, it is traditional to serve this with grits.

  salt fish, flaked
  chicken eggs
  butter
  black pepper
  green onions, finely chopped
  vine-ripened tomatoes
  toast

Soak the fish in water overnight, changing the water a time or two. When you're ready to cook, drain the fish, bone the meat, chop it, and mix it with the eggs in a bowl. Exact measures aren't specified, but I like to have about ½ egg and ½ fish by volume. Stir in a little green onion, including part of the tops. Melt some butter in a frying pan. Scramble the egg-and-fish mixture until done. Add pepper to taste. Serve hot with toast and slices of homegrown tomatoes.

Salt Cod with Parsnips and Egg Sauce

In The Country Kitchen, *a British cookbook, author Jocasta Innes says that this dish was a favorite in medieval times, when salt cod was a way of life. She says the sweet parsnips balance the salty fish. I agree. The measures in this recipe make up a good batch, and you may want to reduce everything by half.*

THE COD

2 pounds salt cod
2 pounds parsnips, peeled and cut into strips
¾ cup milk
⅓ cup butter
pepper to taste
1 teaspoon ground coriander

Freshen the salt fish in several changes of cold water. Then put the fish into a large pan or stove-top Dutch oven and cover with cold water. Bring to a boil, then drain off the water and discard it. Next, cover the fish with a mixture of the milk and ¾ cup water. Add the parsnips. Bring to a boil, reduce the heat to low, cover, and simmer for 45 minutes. Remove the fish carefully and place them in a slow oven to dry. Strain out the parsnips, being sure to retain the stock. Mash the parsnips with a potato masher, adding the pepper and coriander as you go. Stir in a little of the reserved stock until the parsnips are creamy. Serve the fish and mashed parsnips separately, along with the following egg sauce.

THE SAUCE

3 chicken egg yolks, well beaten
¼ cup butter
2 cups reserved fish stock
1 tablespoon flour

Melt the butter in a saucepan. Stir in the flour and then the egg yolks. Slowly add fish stock. Just before serving, remove the saucepan from the heat and stir in the beaten yolk mixture. Stir until the sauce is thick, then turn off the heat and let the sauce coast for a while. Serve the sauce hot over the fish.

International Salt–Fish Specialties

A surprising number of specialty dishes, from Swedish gravlax to Indian Bombay duck, are made with salt fish. Some of these— mostly appetizers—are eaten without being cooked. On first thought, this fact may turn you off from these delicacies, but remember that caviar is not cooked either. I enjoy most of these dishes very much, but I want to make my own, starting with very fresh fish.

Gravlax

In Iceland and Sweden, Atlantic salmon are salted and eaten with a mustard dill sauce. I highly recommend the method for coho or any fresh salmon that you have caught yourself, or that you are certain are very fresh. (This recipe calls for a light salt treatment, which in my opinion should be used only with very fresh fish.) Some books recommend that you use salmon steaks, but boneless fillets work much better. Note also that the name for this delicacy is sometimes spelled gravad lax.

    boneless fillets from a 5- or 6-pound salmon
    3 tablespoons coarse sea salt
    2 tablespoons sugar
    2 teaspoons freshly ground black pepper

2 teaspoons crushed dried juniper berries
fresh dill

Pat the salmon fillets dry with a paper towel and place them, skin side down, on an unpainted board about 6 inches wide and long enough for the fish. (Cut 2 such boards and save 1 for the top.) Mix the salt, sugar, pepper, and juniper berries; sprinkle the mixture over the salmon fillets evenly from one end to the other. Place a few sprigs of dill over one of the fillets, then put the other fillet on top with the skin side up. (In other words, put the salmon halves back together, sandwiching the salt and dill sprigs.) Wrap and overlap the fillets first with wide plastic wrap and then with wide freezer paper. You'll need to seal the fish, remembering that it will be turned over a few times while curing.

Put a plank over the fish and press down on it a little, more or less seating the boards. Then weight the top board with several cans of vegetables (or some other suitable weight of about 5 pounds). Put the whole works in the refrigerator for at least 2 days, turning over the planks and fish every 12 hours or so. Do not unwrap the fish; the idea is to contain the liquid in the package. The salmon should be eaten within 4 days.

To serve the gravlax, remove it from the wrap, drain it, and pat it dry. Put it skin side down and, with a thin, sharp knife, slice it crosswise into thin slices—no more than ⅛ inch thick. Cut down to the skin, then cut the slice away from the skin. This is easy once you get the hang of it. Keeping the salmon cold will make it easier to slice. Serve the salmon slices on a plate with the mustard dill sauce and fresh pumpernickel. Here's what you'll need for the sauce:

¼ cup Dijon or German prepared mustard
1 tablespoon sugar
2 tablespoons white wine vinegar
3 tablespoons olive oil
salt to taste
½ teaspoon ground white pepper
¼ cup minced fresh dill

Thoroughly mix the mustard, sugar, and pepper in a bowl with a whisk. Continue to whisk while adding a little oil in a thin stream. Stop. Whisk in 1 teaspoon of wine vinegar. Whisk, add more oil, whisk, and so on until the oil and vinegar have been used up. Stir in the salt to taste. Then stir in the dill. Transfer the sauce to a serving bowl and refrigerate it until you are ready to eat. The sauce will keep for a few days, but it will need to be fluffed up with a whisk before serving. The same sauce can be used with the following recipe.

Rakørret

This Norwegian dish is usually made these days with farmed rainbow trout. Traditionally, the trout are processed in a wooden container that will hold about 4 gallons, but I have found a Styrofoam ice chest to be satisfactory.

Catch some trout of about 1 pound each and fillet them. Put a layer of coarse salt in the bottom of the container at least 2 inches deep. (You may want to use rock salt for this purpose because sea salt is so expensive these days.) Add a layer of trout, skin side up. Do not overlap the trout. Add a thin layer of salt, covering all the fish but not piling it on. Add another layer of trout, and so on, until you fill the box or run out of fish. Top with salt.

This dish is usually made in the fall, when the weather is cool, and the Norwegians merely sit the box outside in the sun for 3 or 4 weeks. If you live in a hot climate, turn the air conditioner down to 70 degrees and sit the box in a picture window that catches the morning sun. Be warned that this stuff smells almost as loudly as Batarekh (page 27), so keep the lid on tightly if your spouse is hard to live with.

Serve the trout as needed (without cooking) with thin bread and a little hot prepared mustard. If you have prepared a whole box of these trout only to find that you don't like them, or they are too strong or too salty for your taste, soak them in milk overnight to soften and sweeten them. Then dust them with flour and fry 'em in butter. Or flake off the meat and use it to make salt codfish balls, using any good New England recipe. In Boston, according to my copy of the *Old-Time New England Cookbook,* salt codfish balls, Boston brown bread, and Boston baked beans are traditionally served up for breakfast on Sunday morning. Really good Boston baked beans, I might add, are always cooked in a cast-iron pot with a slab of salt pork.

Caviar

If we may believe the critics (and I have no reason to disagree), the best caviar is made from the roe of various sturgeons. Usually, expensive caviar is made from roe with large berries. Part of the gustatory sensation occurs when these berries pop, releasing a burst of flavor. Although caviar can be made from good roe from most fish, such as salmon and carp, I suggest that roe with small berries (or grains) be made into another delicacy, as described in this recipe, or be used for pressed caviar instead. Like winemaking, the processing of caviar can be as complicated and as highly technical as you choose to make it. For openers, try the simple recipe below.

2 cups very fresh roe with large berries
4 cups very good water
1 cup sea salt

Keep the roe on ice until you are ready to proceed—preferably very soon after catching the fish. Find some nylon netting with the mesh a tad larger than the berries. (Netcraft and some mail-order houses carry netting.) Boil the netting to sterilize it, then stretch it over a clean nonmetallic bowl. Break the roe sacs and dump the berries onto the netting, carefully helping them along, and gently rub them about so that the berries drop through the netting. The idea is to separate the eggs from the membrane without breaking the eggs open.

When you have separated the eggs, mix the water and salt, then pour the brine into the bowl over the eggs. Stir gently with a clean wooden spoon, and with your free hand remove any membrane that rises to the top. After 20 minutes, scoop out the berries and place them in a strainer (preferably made of plastic) over another bowl. Put the strainer and its bowl into the refrigerator to drain for an hour or so.

Have ready a sterilized 2-cup jar. (Or several jars if you have landed a sturgeon.) Using a sterilized wooden spoon, put the berries into the jar. The jar should be as full as possible to minimize the amount of air, but the berries should not be broken by packing too tightly. Cap the jar with an airtight lid, then store the caviar in the refrigerator. Note that caviar made by this method is not really cured and should be eaten within 3 or 4 weeks.

I like caviar on thin crackers with a little cheese, but more sophisticated folks have other ideas. In any case, this caviar can also be used in most recipes that call for caviar.

Batarekh

NOSHING FARE FOR THE PHARAOHS

Old salts in Florida and along the Atlantic Coast will be surprised to learn that salt-dried mullet roe was enjoyed by the ancient Egyptians. Ready-made Batarekh is still marketed in Egypt, and possibly in some Paris outlets, where it is called *boutargue* or (I think) *botargo*. European or Egyptian gourmets might argue for the roe of the gray mullet seined from the Mediterranean, or perhaps caught with a hook baited with cooked macaroni, but the weathered folks along the Outer Banks of North Carolina will tell you that the roe of their coastal mullet is the best in the world—far surpassing the golden berries of even the rare sterlet sturgeon. Old crackers on Florida's Gulf Coast will champion their own mullet, some of which are a little different from those that run the Atlantic Coast. Once I could buy fresh mullet roe and milt from my local fish markets, but in recent years, Japan and Taiwan have hogged the market for the sushi trade. They buy the roe by the ton, salt it, dry it, change the name to *karasumi,* and sell it back to us at $70 an ounce, or thereabouts. They also make a similar product, *tarako,* from salt-cured Pacific pollack.

In any case, the roe of cod is also excellent and can be used in this recipe. Menhaden, American shad, hickory shad, mooneye, and other good herring also yield excellent roe. My favorite Batarekh, however, is made from the roe of bluegills, which is readily available to most Americans. Most of the farm ponds in this country are overstocked with bluegills, and these are fat with roe in summer. Catch some and try Batarekh. True, it's an acquired taste, but before long you'll crave more—especially if you have been on a no-salt diet.

Start by carefully removing the roe from the fish, being careful not to puncture or divide the two sacs. Wash the sacs and dry them with a paper towel or soft cloth. Place them on a brown grocery bag and sprinkle them heavily with sea salt. (Ordinary table salt can be used, but sea salt has more minerals and more flavor.) Put the bag in a cool place. After about 2 hours, the salt will have drawn some of the moisture out of the roe and the brown bags will be wet in spots. Put the roe sacs on a new brown bag and sprinkle them again with salt. Change the bag after about 3 hours and resalt the roe. Then wait 4 or 5 hours. And so on until the roe is dry and leaves no moisture on the bag, at which time it will be ready to eat. This will take about 3 days, but after the first 8 hours or so the bag won't have to be changed very often. Small roe won't take 3 days, however, to dry sufficiently. Be warned that Batarekh smells up the house, so it is best to make it on the screened porch or in a well-ventilated place. After it is cured sufficiently, it can be stored for a while in the refrigerator, but it's best to wrap each roe sac separately in plastic wrap. For longer storage, dip each roe sac in melted paraffin.

Beginners should eat Batarekh in thin slices with a drop of lemon juice and a thin little cracker or buttered toast. After the first 2 or 3 slices, omit the lemon and cut the slices a little thicker. Being salty, Batarekh goes down nicely with cold beer. Don't throw out your batch of Batarekh if it doesn't hit the spot, or if you've got squeamish guests scheduled for dinner or cocktails. (As I pointed out in another work, one of my sons won't touch Batarekh, pharaohs notwithstanding, because, he says, it looks like little mummies.) I like to use grated Batarekh as a seasoning and topping for pastas and salads—and it really kicks up an ordinary supermarket pizza. Just grate some on the pizza atop the regular toppings. (I prefer pizza supreme with a little of everything.) Sprinkle shredded cheese over the Batarekh and broil until the cheese begins to brown around the edges and the pizza is heated through. Have plenty of good red wine at hand if you are feeding bibulous guests.

This pizza is one of my favorite quick foods. The grated Batarekh can, of course, also be used to advantage in pizza made from scratch if you are a purist.

Also, I find that grated Batarekh can really pick up a piece of cheese toast for a quick snack. Here are some other ways to use Batarekh—gourmet fare without the mummy image.

Taramasalata

This dish, popular in both Greece and Turkey, is often made these days with smoked cod roe. The original, however, calls for salt-dried roe of the mullet. It's made from tarama, which is similar to Batarekh (page 27) except that the roe is pressed. (The Russians also eat bricks of highly salted and pressed caviar that could be used in this recipe.) Most people won't know the difference.

 3 ounces Batarekh
 4 slices white bread
 1 cup milk
 ¼ cup olive oil
 juice of 1 lemon
 1 clove garlic, crushed
 Greek black olives

Crush the Batarekh in a mortar and pestle until it has a smooth texture. Remove the crust from the bread and soak it in milk. Squeeze the milk out of the bread, then mix the bread into the

Batarekh, along with the garlic. Grind this mixture with the mortar and pestle until the mixture is smooth. Slowly stir in the lemon juice and then the olive oil, tasting as you go. Add more lemon or more garlic if needed to suit your taste. Serve with thin toast or crackers, along with the black olives. I like a little white feta cheese on the side.

Fessih Salad and Pita Bread

According to Claudia Roden's A Book of Middle Eastern Food, *fessih is a fish that has been salted and buried in the hot sand to ripen. When ready, it will be soft and salty. Ms. Roden allows that people in the West may want to use anchovy fillets in lieu of fessih. (Use canned anchovies, or make your own "anchovies"; see page 31.) Also, the recipe requires some tahini, a paste made of sesame seeds. It can be found canned in shops that handle Greek foods.*

    2 2-ounce cans anchovy fillets
    1 large mild onion, sliced
    2 cloves garlic, crushed
    ½ cup tahini
    ½ cup fresh lemon juice
    ½ teaspoon ground cumin
    salt to taste
    chopped fresh parsley
    1 large, ripe tomato, thinly sliced

Put the tahini, lemon juice, onion, garlic, salt, and cumin in a blender or food processor. Zap it until you have a smooth paste, adding a little water if needed. Purists will want to adjust the amount of lemon juice, salt, and cumin until they have an exact flavor and texture. Mince the anchovies and stir them into the tahini cream. Serve in a bowl with large, thin slices of tomato and pita bread. Garnish with parsley.

Anchovies

If you've got real anchovies, you're in business. Also, such finger fish as smelts, shiners, and sand lances will do. If the fish are 4 inches or less, behead and gut them. If they are larger, fillet them. Even small bluegills can be used if they are filleted and perhaps cut into strips to resemble canned anchovies.

Clean and wash the fish according to size. Pat the fish dry. Sterilize some jars with canning lids. Cover the bottom of each jar with a little salt. Add a layer of fish, sprinkling the body cavity of each one with salt as you go. Add a layer of salt, a layer of fish, and so on, packing lightly as you go and ending with a layer of salt. A top space of about ½ inch should be left in each jar. Weight the anchovies with a slightly smaller jar (or other suitable nonmetallic container) that has been filled with water. Put the packed jar (with the weighted jar in place) in a cool place for a week or so. When ready, remove the weighted jar or container. By now, a brine will completely cover the fish. (Fatty fish such as echelon may develop a layer of oil on the surface. Skim this off if you choose.) Seal the jars and store them in the refrigerator. They will keep for a year or longer.

Before serving these in Greek salads and other raw dishes that call for canned anchovies, I like to remove the fish from the brine, rinse them, pat them dry, and dip them in olive oil. Note that the salt will have softened the small bones in the fillets. Whole fish should be boned, which is easily accomplished by spreading the body cavity, pulling off one side, and lifting out the backbone.

Salt Salmon Birdseye

According to George Leonard Herter, the late Clarence Birdseye, the father of frozen supermarket foods, came up with the following recipe for salmon and other fatty fish, such as bullheads. Further (Herter says), the same dish was once called salmon fuma in New York City.

> 3 pounds skinless fresh salmon fillets
> 1 gallon water
> 2½ cups salt
> 5 teaspoons Liquid Smoke
> ⅙ ounce sodium nitrate
> 1/32 ounce sodium nitrite

Pour the gallon of water into a crock or other large nonmetallic container. Dissolve the salt in the water, along with the sodium nitrate and sodium nitrite. See whether a chicken egg will float in the solution. If not, add more salt until the egg rises. Then stir in the Liquid Smoke. Put the fillets into the solution and weight with a plate or some nonmetallic object of suitable size; the idea is to keep the fish fillets completely submerged at all times. Leave the fish in the solution for 4 days. (If you feel compelled to stir the fish a time or two, use a wooden spoon.)

Drain the fish. Before serving, slice the fillets into very thin slices. Partly freezing the fillets and slicing with a sharp, thin knife will help. Serve atop crackers.

AIR-DRIED FISH

Although air-dried fish—one of man's first preserved foods—is very important from a historical viewpoint, the method has not been popular in America in recent history and has not been widely practiced except by some native peoples in Alaska and Canada. Fish has often been dried for use as food for sled dogs, partly because drying reduces the weight by about 80 percent. In parts of Africa and no doubt other places, fish were dried with the aid of smoke. The smoke helped keep the blowflies away from the fresh fish, and was sometimes discontinued after a day or two.

Stockfish

In the recent past, very large quantities of dried fish were produced commercially in Scandinavia and Iceland, where the dry air and cool breeze made the process feasible on a large scale. Air-dried cod—called stockfish, from the Norwegian *stokkfisk* or Swedish *torrfisk*—provided the people of the Middle Ages with food, and was later shipped in large quantities to parts of Africa, where dried fish were preferred to salt fish (and still are in some places). Although refrigeration has hurt the stockfish trade, tons of cod are still air-dried commercially in Scandinavia each fall; Norway alone exports 50 million to 55 million pounds of stockfish to Africa in a year, according to A. J. McClane's *Encyclopedia of Fish Cookery*.

Typically, those cod destined to be stockfish are merely gutted and hung out to dry on huge wooden racks. Of course, a few fish can be dried for home consumption without large racks; often, the gutted fish are hung under the eave of the house. The drying will take from 2 to 6 weeks, or even longer, depending on the weather and the size of the fish.

The dried fish become very hard, and they require lengthy soaking in fresh water before they become suitable for human consumption. A medieval recipe calls for boiling dried fish in ale, then shredding them and mixing in shredded dates, pears, and almonds. The mixture is reduced to a paste in a mortar and pestle, then shaped into patties, dusted with flour, dipped in a batter, and fried in hot oil.

Modern practitioners may prefer the following recipe.

Traditional Finnish Christmas Fish

Several weeks before Christmas, the dried fish are put into a solution of 1 gallon of water and 1 tablespoon of lye; then they are soaked for 2 weeks. For another week, the fish are soaked in fresh water, changed daily, to leach out the lye. Then the fish are poached gently until the meat flakes easily. The flaked meat is

Fish hanging to dry under the overhanging roof of a house

put into a white cream sauce seasoned only with freshly ground black pepper, and served with boiled potatoes.

Racking

Racking is used to dry rather large fish with a low fat content, such as cod, hake, or even flounder. The fish are cut in half, then the backbone, rib bones, and head are removed, leaving the collarbone intact. The sides of the fish are cut into long strips about 1 inch wide; these are left joined together at the collarbone. After being washed, the fish are soaked for an hour in a saturated salt brine (that is, a brine with enough salt to float an egg). They are then hung by the collarbone in a dry place, out of direct sunlight. Drying takes from 1 to 2 weeks, after which the fish can be stored for future use.

The dried fish are soaked in fresh water to freshen the flesh, then creamed or flaked and used in recipes for chowders, fish loafs, or fish cakes. At one time, the dried strips were eaten without cooking, like jerky.

Mexican Sun-Dried Shark

According to A. J. McClane's *Encyclopedia of Fish Cookery,* a major shark fishery exists in Mexico's Sea of Cortés. In fact, the industry is centered on Isla Tiburón, or Shark Island. The sharks include the mako, brown, blacktip, hammerhead, tiger, bull, leopard, nurse, thresher, and horn. After processing, most of the meat, McClane says, is sold in Mexican markets as salt cod. The fins are dried and sold for Chinese sharkfin soup. The back meat is cut into fillets and soaked for 20 hours in a weak brine of 4 pounds of salt to 10 gallons of water. This soaking leaches out the uremic acid, which is present in most sharks and which gives the meat a strong smell of ammonia.

After brining, the shark fillets are dried in the sun. McClane says that careful drying is critical in preservation, and that the fillets must be evenly exposed to the sun on both sides and protected from damp night air.

Dried Shrimp

For this method of drying shrimp, I am indebted to Frank G. Ashbrook, author of *Butchering, Processing and Preservation of Meat.* Although shrimp of any size can be dried, the smaller ones work best and are less desirable for the market or other methods of home use. People who have tried to peel enough tiny shrimp to stay ahead of their appetite will see the advantage!

For best results, start with very fresh shrimp. Wash these and bring to a quick boil in salted water, using 1 cup per ½ gallon of water. (Do not try to boil too many shrimp at the same time because they will lower the temperature of the water.) Boil the shrimp for 5 to 10 minutes, depending on size. Drain the shrimp and spread them in the sun to dry. If bugs are a problem, rig a way to cover the shrimp with a fine-meshed screen. The shrimp can be rather crowded, but should not form a layer more than 1 inch deep.

For the first day, the shrimp should be turned every half hour. At night or during a rain, remove them to a dry, well-ventilated place; do not merely cover the shrimp with a tarpaulin or other direct covering. If you are really into drying, it's best to build movable trays with wire or slat bottoms; then the whole tray can be taken inside at night or during a rain. In most areas, you'll also need a screen cover to keep the flies off the shrimp, or perhaps to keep the seagulls away.

Drying small shrimp will require 3 days in sunny weather, longer if the days are short of sun and wind. When the shrimp are dry and hard, place them in a cloth sack. Beat the sack with a board. This will break the shells. Then winnow the shrimp in a sifting box, made with a wooden frame and ¼-inch mesh. The bits of shell will fall through, leaving the dry shrimp meat on top. This will be much smaller and lighter than the original. In fact, 100 pounds of shrimp will shrink down to 12 pounds of meat. The dried shrimp can be put into jars and stored in a dry place.

The dried shrimp can be used in soups and stews that will be cooked for some time, or they can be freshened by soaking in water for several hours. Freshened shrimp can be dusted with flour and fried or sautéed in butter, or they can be eaten raw as appetizers.

Asian and regional Mexican cuisines make good use of dried shrimp, which can be purchased in some ethnic markets. Usually, the shrimp are dried with the aid of salt. In addition to shrimp, the Chinese also dry and market oysters, squid, sea cucumbers, and even jellyfish. Scallops are sometimes air-dried.

Easy Dried Shrimp

Here's a technique that I like to use for quick-drying 2 or 3 pounds of shrimp. Peel and devein the shrimp, place them in a shallow tray, and cover them with a cure made with 1 pound salt, 2 cups brown sugar, 1 tablespoon onion powder, and 1 teaspoon ground allspice. Place the tray in the refrigerator for 6 or 8 hours.

Wash the salt cure off the shrimp and pat them dry with paper towels. Place the shrimp on cookie sheets and dry in the oven for about 12 hours on low heat. Use the lowest setting on your oven, and leave the door ajar. After drying the shrimp, pack them into airtight containers and refrigerate them for up to 3 months, or freeze them for longer storage. Use the shrimp in soups and stews, or in any Asian recipe that calls for dried shrimp.

Air-Dried Salt Fish

One purpose of adding lots of salt to fish is that it draws out the moisture, thereby speeding up the drying and curing processes. Usually, small fish to be salt-dried are gutted and beheaded; larger ones are filleted, leaving the collarbone intact to help hold the fish together while hanging.

After dressing, the fish are usually washed in salted water, made by adding 1 cup of salt to each gallon of water. The fish are drained, then dredged in a box of salt. Another box is lined on the bottom with salt, and the dredged fish are put down in a layer. Salt is scattered over the layer, then another layer of fish is put down, and so on. As a rule, the total amount of salt used in this process is 1 pound per 4 pounds of fish. Using too much salt will "burn," or discolor, the fish.

The fish are left in the salt for at least a day, or up to a week, depending on the size of the fish and the weather. Then the fish are rinsed and scrubbed to remove the brine. Next, they are drained for a few minutes and hung on racks or placed on drying trays. They are usually kept in a shady place with a good breeze.

Fish Jerky and Pemmican

In suitable climates, fish can be cut into strips and dried like jerky, in the sun or wind. They can also be dried atop large rocks,

provided that they are turned from time to time; the rocks usually hold heat from the sun and speed the process considerably,
especially if the rocks are exposed to the sea breeze but out of
the spray.

I recommend that the fish strips be soaked in brine for an
hour or so before drying, but this is not necessary in all cases.
I also recommend that only lean fish be dried in this manner.
Once dried, the fish jerky can be eaten as is or used in stews. It
can also be pulverized and used in pemmican.

Jerky and pemmican, which are usually made with red meats,
are discussed further at the end of Chapter 3.

3

DRY-CURED MEATS

I consider the highly advertised Virginia hams to border on being fraudulent. There has never been a time that they could measure up to the well turned out Tennessee or Kentucky ham.
—Madison Ames Saunders Jr., *A Man's Taste*

Meats cured with dry salt rather than brine are covered in this chapter; Chapter 4 covers brine-pickled meats.

Although beef and other red meats can be dry-cured, in this country and in most other parts of the world pork is by far the most popular salt-cured meat, as in salt pork, bacon, and ham. At one time, pigs were raised in towns and cities as well as on the farm; in fact, the pigs were allowed to roam the streets, thereby helping control the garbage problem instead of adding to it. I have raised a pig or two in a pen inside town limits, but these days you really need a place in the country to raise your own. If you buy fresh pork for curing from a local farmer or a meat processor, make sure that you get it very fresh and properly chilled.

Dry-curing red meats and making jerky are covered later in this chapter.

SALT PORK

Salt pork is one of the world's great seasoning meats and is often used in such dishes as Boston baked beans and hoppin' John as well as in fish chowders. It can be made from side meat or from fatback. The jowls are also used, and these are traditionally eaten on New Year's Day in some areas. I prefer to use a good cut of side meat so that it will have a lot of lean meat along with the fat. In fact, I consider the whole side of pork to be salt pork if it is cured and not smoked; if smoked, it's bacon, I say. Usually, salt pork is the thickness of slab bacon and is prepared with the skin on. Slabs can be cut into convenient lengths and widths. I find that 6-by-12-inch pieces work just fine. You can salt-cure the whole side, then trim it for storage; of course, the trimmings can be used for seasoning meat.

½ side of hog meat, trimmed and cut into sections
lots of salt

Rub the pieces of side meat on all sides with salt. Place each piece on a bed of salt, preferably on a wooden plank or in a shallow wooden box or tray, then cover it with salt. Place the meat in a cool place for 1 week. Resalt the pork and leave it in a cool place for another week; after that, it can be cooked as needed.

Although cured salt pork can be packed in boxes and barrels, it will sometimes become rancid, especially in warm weather. It's best to wrap each piece separately and store it in the refrigerator or freezer until needed.

Note: If you prefer, add some sodium nitrite to the salt that is rubbed onto the meat. Mix 1 ounce of sodium nitrite with each pound of salt. It is important that the sodium nitrite be mixed well with the salt so that it is distributed evenly.

Salt Pork Breakfast

These days we are constantly told that animal fat is not good for us. If you tend to eat too much of a good thing and have no self-control, or if you are under a doctor's orders, proceed at your peril.

salt pork with lean streaks
cornmeal
oil
2 tablespoons all-purpose flour
2 cups milk
black pepper
hot biscuits

Slice the salt pork to about ⅛ inch thick and cut off the rind. Allow at least 4 slices per person. Put the slices into a pan and cover them with water. Bring to a quick boil, cover, and simmer for 10 minutes. Drain the salt pork and discard the water. Heat about ⅛ inch of oil in a skillet. Coat the salt pork with cornmeal and fry it in hot fat until brown and crispy. Put the salt pork pieces on a brown paper bag to drain.

For gravy, pour all the grease out of the frying pan except for about 2 tablespoonfuls. Heat this grease and then stir in the flour. Using a wooden spoon, stir well until the flour is brown. Slowly add the milk. Add a little black pepper to taste. Stir and simmer until the gravy has the consistency you want. Serve the gravy over hot biscuit halves and eat with the fried salt pork strips. Have plenty of hot coffee. This dish also goes nicely with cold slices of homegrown tomatoes.

Yankee Baked Beans

Beans were raised by Native Americans and, along with corn, formed a big part of the early colonists' diet. Although there are hundreds of variations on baked bean dishes, often called Boston baked beans, here is one that I highly recommend. Most of the Boston recipes seem to call for molasses, which was shipped up from Jamaica for making rum; other recipes call for brown sugar. Purists might even hold out for maple sugar.

1 quart dried navy beans
½ pound slab salt pork
½ cup molasses
1 medium onion
½ tablespoon salt
½ teaspoon powdered mustard

Put the beans into a nonmetallic container, cover them with water, and soak them overnight. Drain the beans and put them into a cast-iron pot. Cover with fresh water, bring to a boil, and simmer for 1 hour. Drain the beans.

Preheat the oven to 250°F and put on some water to boil. Chop the onion and put it into the bottom of a large cast-iron pot. Add the beans to the pot. Mix the molasses, mustard, and salt, then spread this mixture over the beans. Put the slab of salt pork on top of the beans so that the rind is up. Cover the beans and pork with boiling water. Put the lid on the pot and bake for 8 hours. Add a little water from time to time if needed, but do not stir, leaving the slab of pork on top.

COUNTRY HAMS

A real country ham is salt-cured and aged. It may also be smoked, but not necessarily so, and it may have some sugar in the cure, in which case it can be called sugar-cured. Most modern curing recipes recommend that you add sodium nitrate or sodium nitrite, or both, and others recommend saltpeter, Prague Powder, and such formulas. Some texts even advise you to buy an injection pump, and offer illustrations on how to use this device. Unless you're highly experienced, you'll need an X-ray machine to inject the ham in the right places. Moreover, if you are going to do all this and want to pump your ham full of liquid, you might as well buy one that an expert has "cured" for you.

If you want what I consider to be the real thing, forget the injection pumps and all the cures except salt. Then proceed with great care and caution, getting the various steps right. There is no shortcut. But before setting forth a ten-step recipe in detail, I would like to quote another man's recipe and comments. Although I might frown at his use of Liquid Smoke, I really can't argue with the results. What I really like is the man's spirit.

Saunders's City-Cured Country Ham

A culinary sport by the name of Madison Ames Saunders Jr. has formed some firm opinions about country hams, and his comments are in line with my experience and conclusions. The following excerpt is from *A Man's Taste*, published by The Junior League of Memphis.

You might think that it is difficult to cure a country ham while living in the city. Not so. When I became interested in curing a country ham, I read everything I could find including a Department of Agriculture pamphlet on the subject. This last was a mistake. True to form, the pamphlet only served to confuse. By this time, though, I realized that hams had been cured for hundreds of years, often by illiterate people, as a method of preserving pork. If illiterate people can do it, I thought, why can't I? Armed with all of my new-found knowledge, I then sought out my good and great friend, George F. Jackson. George spent a great deal of his youth following a pair of mules on small farms in northern Mississippi, northern Shelby County, Tennessee, and on one located on Centennial Island in the Mississippi River. He is incredibly wise in all things pertaining to the country, and I was to fill in the missing portions of information by questioning him.

By this time it was the middle of November, the traditional time to cure meat in this part of the country. First I built a plywood salt box. Size doesn't make too much difference, but mine is roughly 4 feet by 5 feet and is 1½ feet deep. I also drilled 8 or 10½-inch holes in the bottom and built a rather snug-fitting plywood top.

You must use non-iodized salt in the curing process, and this may require a trip out in the country. Any store that caters to the farm trade will have it, and at a very reasonable price. It comes in 25-, 50-, and 100-pound bags so don't stint. While you are there, also buy a quart of Liquid Smoke. It will save you a trip later on.

Now you are about to get into business. Go to a packing house or your friendly butcher and buy 4 20-pound fresh pork hams. I like 20-pound hams because they will lose 5 to 6 pounds in the curing process, and a 14-pound ham is about the best size. . . .

Put your salt box on the 4 concrete blocks in your garage or carport and cover the bottom of the box with 2½ to 3 inches of salt. Nestle the hams, skin side down, in the salt and then cover with salt so that no piece of ham is not in contact with it. Put the cover on the top and put 2 concrete blocks on the cover. This is to discourage city varmints such as cats, rats, dogs, raccoons, etc., that might, in the silence of the night, want to become silent partners. At least once a week remove the cover to the box and peer at the mounds of white salt. Don't disturb the salt or poke around. This weekly observation doesn't help the ham, but it increases your pride in accomplishment and anticipation.

In about 4 weeks the hams are ready to be removed from the salt. If the weather has been extremely cold, 15 degrees or below, cover the box with a blanket or an old rug to keep from freezing. If the weather remains extremely cold, leave the hams in the salt for 5 to 6 weeks, instead of the recommended 4. Remove the hams, brush the salt off, and paint with liquid smoke. Do this 3 nights in a row, returning the hams to the salt box for safe keeping but no longer covering them with salt.

Now they are ready to hang. Wrap each ham with about 3 layers of cotton cloth, old bed sheets, old undershirts, or anything that you think will keep insects from getting to the ham. I hang mine in my debugged basement. It's not a bad idea to hang a Shell No-Pest Strip close by.

Now we play the waiting game. The hams will be ready to eat in 6 months, but I believe that they really reach their peak when they have been hanging for 2 years. In the meantime, pinch, feel, fondle, and smell at regular intervals. Like other things I know of, anticipation is sometimes as good as realization.

At last comes the great day. When the hams have been hanging for 6 months, cook one using my recipe [page 55]. After you have tasted it, invite your neighbors in for a drink. Put the whole ham out, with a very sharp knife for their enjoyment. Now comes the difficult part. You must be modest. As they rave about this culinary delight extraordinary, don't tell them that you both cooked and cured it. Let your wife tell them. When they stare at you in amazed disbelief that a mere mortal can do this, and start a barrage of questions, shyly admit that you have more aging in the basement, claiming all the while that it really wasn't so much.

A. D.'s 10-Step Georgia Ham

As I hope has been made clear, there is far too much conflicting information about cured hams. One authority might advise you to use ascorbic acid instead of sodium nitrite or sodium nitrate, and another will tell you not to substitute. Others will say that neither ingredient is necessary for curing hams. One authority says to age the hams before smoking; another, after smoking. Some writers will say that country hams are best when they are aged for a year or longer; others say that they get too hard after 6 months. One authority says to sprinkle pepper over your hams after curing to prevent mold; another says that pepper causes mold. In short, writers and practitioners, along with booklets written in Federal Prose, have confounded the issue, and trade books on the subject have tried to treat cold-smoking and hot-smoking (*i.e.*, cooking) in the same work, thereby causing even more confusion—and more margin for errors of serious consequence.

Guidance is needed, but not in the form of magic cures and gadgets. Those people who think that the problems can be solved with Prague Powder and brine injection pumps are wrong. For one thing, hams are more exacting to cure because they are large. Yet, owing to complicated enzymatic chemistry, they can be the most rewarding of the cured meats. Anyone who has sat down to eat a properly cured and expertly cooked country ham needs no further reason to proceed.

The process is really not difficult. Yet things can go wrong. My older brother, for example, once had a small farm on the Choctawhatchee River, where he let his hogs forage on acorns and rootables. Starting in late summer, he fattened some prime pigs on corn and peanuts; then, on first frost, he butchered a few for home use. One fall, he butchered four prime pigs and salt-cured the hams and shoulders, stacking them on a wooden shelf in his smokehouse. The shoulders worked out just right, but the larger hams soured and had to be thrown out. Why? He didn't know. He thought he had followed the exact procedure that he had always used, and the one that my father and grandfather had used before him.

In any case, I consider the following steps to a country ham to be more important than brine pumps and secret formulas.

1. Select a good hog—not too fat and not too lean. About 200 pounds on the hoof will be just right.

2. Scratch the hog down with a corncob until it starts to snore, then dispatch it quickly. (At least, don't chase the hog or otherwise rile it up before the

slaughter.) Some people will want to bleed the hog, but remember that merely cutting its throat and jugular vein won't accomplish much except cut off the blood to the brain; for bleeding the meat, the hog should be "stuck" in order to sever the aorta artery.

3. Quickly gut the hog and scald it, then scrape off the hair in a vat full of boiling water. This process leaves the skin on the hams and other parts.

4. Chill the carcass as soon as possible. The carcass should cool down to at least 40°F through and through. If you have access to a mechanical cooler, use it. If not, you'll have to pick your time for killing hogs. On a cold day, perform step 3 in the afternoon and hang the hog in the night air to chill. A night with a light freeze is ideal.

5. When the meat has cooled down, reduce the carcass to hams, shoulders, etc. (This can be done in step 4, but it's much easier if you chill the meat before cutting it; cold meat is simply easier to cut and handle.)

6. Trim the visible fat from your ham, but leave the skin intact.

7. Move the ham to a salting table or bench and apply the salt cure. Use at least 8 pounds of salt per 100 pounds of ham. Salt alone will cure the ham, but

spices and other ingredients can also be mixed in for one purpose or another. Sodium nitrate, sodium nitrite, Prague Powder (a commercial curing mixture), and saltpeter preserve and enhance the bright red color of the ham, but they have little curing power. Sugar helps the flavor, the surface color, and the texture of a ham (helping prevent it from becoming too hard). Prague Powder is available in various formulas, some designed to give continuous aid to the curing process. Prague Powder 2, for example, is essentially a mixture of salt, sodium nitrite, and sodium nitrate. The thinking here is that the sodium nitrite is used up fast and the sodium nitrate is used up slowly, thereby giving continuous action. Both sodium nitrite and sodium nitrate are toxic when used in large amounts, and state and federal governments have set limits on its use in commercial hams. (I repeat that neither cure is necessary for home use; their purpose is primarily to give the cured ham a red color.) Saltpeter is used in some of the old cures; it may not be a safe chemical and should be used only in small amounts.

If you want a good general cure for hams, I recommend the 10-1-2-1 formula. That is, 10 pounds salt, 1 pound sugar, 2 ounces sodium nitrite, and 1 ounce sodium nitrate. Dry spices and herbs can be added to the mix, but I don't think they are necessary. It's hard to beat the flavor of ham.

It's best to apply the cure (or plain salt) at three different times during the curing process. Apply ⅓ on the first application; ⅓ after 7 days; and ⅓ after 17 days. For a 16-pound ham, use 8 ounces of cure per rub; for an 18-pound ham, 9 ounces; 20-pound, 10 ounces. (These figures are from "Curing Georgia Hams Country Style," a booklet published by the University of Georgia College of Agriculture.) Rub the cure all over the ham, paying particular attention to the bones on either end. It's a good idea to pack a little salt or cure into any opening around the bone.

8. After the first rub, place the ham on a shelf in a cold place, preferably at about 36°F. (If the temperature is colder, increase the length of the cure a few days.) Do not cure the ham at temperatures above 40°F, because the "ham-souring" bacteria can multiply and cause the meat to spoil.

   After 7 days, rub the ham with the second application of the cure (⅓ of the total). After 17 days, rub the ham with the final ⅓ of the cure. Then complete the curing period, the length of which depends on the thickness of the ham. These figures (which I consider to be about right) are given by the aforementioned Georgia publication: 4 to 5 inches of thickness, 28 to 35 days; 5 to 6 inches, 35 to 42 days; 6 to 7 inches, 42 to 49 days. The reason for the long curing time is that the salt penetration is slow; the thicker the ham, the longer it takes. (The idea behind brine pumps, of course, is to speed up the salt penetration. If used by a skilled operator, they might do that, but the danger is that all areas of the ham may not be injected properly. I don't recommend their use, partly because I don't want water and chemicals pumped into my ham.)

Note that the ham will shrink considerably during the curing period.

9. At the end of the curing period, wash the excess salt off the ham, dry it, and put it in a cool place, or better, keep it under refrigeration at 40°F. Leave it for 20 days. Why? So that the salt will penetrate throughout the ham. After the curing period, most of the salt will be near the surface; after 20 days, it will be more evenly distributed (but not exactly even) throughout the ham.

Called salt equalization, this process is essential for a properly cured ham. Most of the spoilage in country hams is caused by inadequate salt equalization and occurs during the aging process.

10. After salt equalization, hang the ham in a cool, airy place for aging. It can be hung by the shank end with cotton string, or it can be put into a stocking made of net material. (These are available from supply houses.) Wrapping the ham in clean sheeting will be fine. Air circulation is important during aging; consequently, the ham should never be wrapped in airtight material.

If all these steps have been duly observed, the temperature of the aging environment isn't as critical as for curing and salt equalization. According to the Georgia booklet, "When no controlled conditions are available, hams age best during the summer months as the inherent enzymes that produce the aged flavor are more active." The aging temperature should not exceed 95°F, however. During aging, the ham will lose more of its weight and shrink in size.

For how long should the ham be aged? Six months is a reasonable minimum figure, but many country hams are kept and treasured for generations. I worked with a student of nuclear physics at Oak Ridge, Tennessee, who boasted that his old East Tennessee family had treasured hams that were a hundred years old.

In any case, a country ham may develop mold on the surface during aging. This is completely normal, and the mold is easily washed off before cooking; for this purpose it's best to use a stiff brush and warm water. Remember that salt-cured hams are much firmer than today's supermarket "cured" hams. They require soaking in water and longer cooking times. Often, they are simmered in water before being baked. Recipes for cooking country hams are covered later in this chapter.

Smithfield (or Virginia) Ham

I once tuned in on a TV documentary about Smithfield hams, in which they showed huge buildings, like barns, full of curing hams. Although I didn't learn any trade secrets, I saw enough to know that this is a highly controlled and exacting process that, like making a certain wine, cannot be duplicated easily at home; part of the problem, it seems to me, is making a product that is consistent from ham to ham, batch to batch, year to year. Sometimes, these Smithfield hams are touted as being fed or fattened on peanuts, but I think the secret is in the curing. Also, the name

Smithfield is probably exploited commercially. In any case, my information on the cure is based almost entirely on Frank G. Ashbrook's *Butchering, Processing and Preservation of Meat*:

> These hams are cut with the long shank attached. They are cured in a dry mixture for 5 to 7 days, depending on their weight. They are then overhauled, resalted, and held in cure from 25 to 30 days (1½ days per pound). After this dry cure is completed, the hams are washed in warm water, dried, sprinkled with pepper, and cold-smoked (at 70 to 90 degrees) for 10 to 15 days, after which they are aged and mellowed by hanging in a dry room. These hams improve with age and are in perfect condition when 1 year old.

Note that the hams are cold-smoked for 10 to 15 days according to Ashbrook, and are then aged. The Georgia booklet says to smoke the hams *after* the aging process. From a safety viewpoint, it probably doesn't matter as long as both the salting and the salt equalization steps have been successful.

Cooking a Dry-Cured Country Ham

Most people don't know how to cook a country ham. Some of the old cookbooks don't go into much detail; no doubt those authors assumed that any fool would know to freshen the salty ham before cooking it. In any case, some country hams end up being not only too salty for human consumption but also too hard and too tough to eat. Of course, a cured country ham is supposed to be firmer than a modern embalmed ham, and as a rule it should be carved in thin slices. In fact, some of the world's more famous cured hams are eaten raw and part of the secret is very thin slices.

In Tennessee, I once ate in a restaurant that had quite a local reputation for country hams. The ham was simply machine sliced and pan-fried. Well, it was edible, but it was too hard and too salty to suit me. The redeye gravy, however, was just right for sopping with biscuits. The owner of the restaurant told me, on my next visit, that I had to buy ham in order to get the gravy. I did so.

Saunders's Boiled Country Ham

Weigh the cured ham, then place it skin side down in a large, deep pot. Add enough water to cover the ham. Then add 1 cup vinegar, 1 quart molasses or sorghum, 1 cup powdered instant coffee, and 6 or 7 dozen whole cloves. Bring the liquid to a boil, cover the pot, reduce the heat, and simmer for 20 minutes for each pound of meat. Turn off the heat and allow the ham to cool overnight in the liquid. Remove the cooled ham from the liquid and place it skin side down on a thick section of newspaper. Refrigerate the ham for 24 hours. Trim off most of the skin and fat. Now you're ready to slice and eat the ham.

A. D.'s Whole-Ham Favorite

Here's an old three-step method of cooking a country ham that combines soaking, boiling, and baking. It's hard to beat, making a beautiful ham as well as a tasty one without too much salt. This recipe can be used for a whole ham or just a butt or shank portion. Before proceeding, make sure you have large enough containers for holding and cooking the meat.

Step 1

cured country ham
a little baking soda

Scour the ham with water and a brush dipped into baking soda. Put the ham into a large container such as an ice chest and cover it well with cool water. Let it soak all night, changing the water a time or two.

Step 2

soaked ham
1 cup wine vinegar
1 stick cinnamon
20 black peppercorns

The next day, put the soaked ham into a large pot or container and cover it with fresh water. (If you are cooking a whole ham and don't have a container large enough or deep enough, consider using one of the large patio fish fryers that fit in a rack over a bottle gas cooker with double burners; a tall stockpot can also be used.) Bring the water to a boil, add vinegar, cinnamon, and peppercorns, then quickly reduce the heat and simmer (or poach) the ham for 30 minutes per pound. Add a little more water from time to time if needed. Remove the ham and let it cool in the pan liquid.

Step 3

1 soaked, poached, and cooled ham
1 cup brown sugar
1 cup fine bread crumbs
1 cup prepared mustard
whole cloves

Preheat the oven to 325°F. Skin the ham, leaving the fat on the meat, and score the top surface with a small knife; a 1-inch diamond pattern is ideal. Put the ham into a roaster or onto a suitable baking pan. Combine the mustard, bread crumbs, and brown sugar, then cover the ham with this mixture. Stick a clove in the center of each diamond. Put the ham in the oven and cook for 30 minutes, or until nicely browned. Small hams and portions of hams may take less time to brown; large whole hams may take a little longer.

Fried Country Ham and Redeye Gravy

Here's a hearty breakfast dish that can be made with leftover baked country ham or with center-cut slices of country ham. Some people make the dish without freshening the meat, resulting in a piece of ham that is too tough and salty for good eating. The gravy, however, will be good. Many Southerners insist on serving redeye gravy with grits, but it is even better over fluffy white biscuit halves.

    ham slices about ¼ inch thick
    lard or bacon drippings
    ½ cup black coffee
    black pepper
    hot biscuits

Heat a little lard or some bacon drippings in a cast-iron skillet. Fry the ham slices for 5 or 6 minutes on each side, then place the slices on a serving platter. With a wooden spoon, scrape up the pan dredgings. Add the coffee, increase the heat, and simmer for 5 minutes. Season with pepper. Serve the gravy over the ham and biscuit halves.

Note: This recipe makes a thin gravy. It can be thickened with flour or perhaps with cream, and it will be very good, but it won't be redeye gravy.

EUROPEAN DRY-CURED HAMS

Though it is somewhat difficult to decide which one of the various kinds of ham should be adopted, in my opinion that of Bohemia, known as Prague Ham, is best for a warm dish, and that of York for a cold dish.

—A. Escoffier

I like to think that the perfect ham—salt-cured and hickory-smoked—was developed in the Americas, but this may not be the case. According to my old edition of *Larousse Gastronomique*, it was the Gauls who developed the technique of curing hams. After salting the hams, the Gauls smoked them for 2 days with "certain selected woods." Then they rubbed them with oil and vinegar and hung them to age. The Gauls are said to have exported these hams in large numbers to Rome and to the whole of Italy. But I'm sure that champions of Italy's Parma ham, which is fed on the whey left from making Parmesan cheese and fattened on parsnips, would have a few words to say about the origin of cured hams, so I won't pursue the matter further.

Prosciutto

This delicious Italian ham is salted and air-dried. It is often served raw, thinly sliced, as an appetizer.

1 fresh ham with skin
4 pounds coarse salt
1 pound sugar
¼ pound fresh pork fat
2 cups red wine
2 tablespoons minced garlic
1 tablespoon flour
1 teaspoon saltpeter
salt and freshly ground pepper

Mix a dry cure with the coarse salt, sugar, and saltpeter. Weigh the ham, then put it onto a wooden worktable or countertop. Rub it well with the salt cure, packing the cure around the bone at either end. Place the ham in a wooden box and store it in a cool place (between 36°F and 40°F) for 2 days for each pound of ham. Check the ham every 7 days. If the salt has melted and run off, resalt the ham.

At the end of the curing time, remove the ham. Rinse it under running cold water, then rinse it with wine. Tie a cord to the shank end and hang the ham in a cool place for 2 days.

After 2 days, mash some garlic into a paste and rub the exposed meat of the ham. Dice the pork fat, put it into a skillet, and fry until fat is rendered. (Save the residue for crackling bread, or eat the bits on the spot.) Make a paste of the pork fat, flour, salt, pepper, and about 3 tablespoons water. Cover the ham with this paste. Wrap a sheet of cheesecloth around the ham, then place it in a large brown paper bag. Tie the paper bag loosely and hang the ham in a cool, airy place for at least 3 months.

Ibérico Ham

Jamon Ibérico is a favorite noshing food in Spain, home of the tapas, or "little dishes." There are several of these hams, all quite expensive. For example, a 16- to 18-pound jamon serrano by Fermin sells for $795. The key to gourmet Spanish hams is in range-fed pigs, free to roam the forests in search of acorns and other wild edibles. The hams from these pigs are salted and cured in cool mountain air.

The Spanish gourmets usually eat these hams raw, thinly sliced, along with various cheeses, olives stuffed with anchovies, beautiful roasted red piouillo peppers, green Guindills finger peppers, white asparagus spears, tiny artichoke hearts, almonds, dry-cured chorizo sausages (made from the same forest-fed pigs), and other finger foods. I like to add a log or two of palm hearts, some pork cracklings, melon balls, and quail eggs.

I go on at some length about these Spanish goodies because the feral pig in North America, which has roamed free since the Spanish introduced the hog to Florida and Mexico, enjoys pretty much the same diet as the Spanish free-range pig. The feral American pig, sometimes called the razorback or piney woods rooter, now grows in the wild across the country and is considered a nuisance in some areas. This affords a rare culinary opportunity for American hunters who are also culinary sports. The successful pig hunter should butcher the pig as soon as possible, being sure to scald the carcass and scrape off the hairs instead of skinning it. Salt and cure the two hams and two sides of bacon. Grind the rest of the meat and make chorizo or other sausage (for how-to details, see my book *Sausage*).

OTHER SALT-CURED MEATS

Although pork is the most popular meat for salting, other meats can also be used. In Russia, bear hams are cured exactly like pork hams. In many parts of the world, salt mutton is popular. In almost all cases, it's best to get the meat dressed and chilled as soon as possible, then apply salt as soon as the meat is chilled.

Although opossums, rabbits, and other small animals and game can be salt-cured, it is usually larger animals and larger cuts of meat that are preserved.

Missouri Venison Cure

Here's a recipe that I found in Cy Littlebee's Guide to Cooking Fish & Game. *It was reprinted there from a 1954 edition of* Missouri Conservationist, *to which it was submitted by Don Baggs, a sport from St. Louis. I tried it with buffalo and found it to be very good.*

    3 pounds salt
    4 tablespoons ground allspice
    3 tablespoons black pepper

As soon after killing as possible, dissect the thigh, muscle by muscle. Skin off all membranes so the mixture will contact the raw, moist flesh. Best size for the pieces is not over 1 foot long by 6 or 8 inches wide and 4 inches thick.

Rub on the powder thoroughly. Then hang up each piece of meat by a string in the small end and let it dry in the wind. If the sun is hot, keep the meat in the shade. (In the North, the sun helps the process.) Never let the meat get wet. If the weather is rainy, hang the meat rack by the heat of the campfire. Don't let it get any more smoke than necessary and cover with canvas at night.

Meat prepared like this is not at its best until it's about a month old. After that, no hunter or trapper can get enough.

Pueblo Venison Cure

This recipe contains some ingredients brought to the Southwest by the Spanish, and the original probably contained red pepper instead of black. The allspice is, however, a New World ingredient.

> fresh venison, in chunks
> 3 pounds salt
> 5 tablespoons black pepper
> 4 tablespoons ground cinnamon
> 4 tablespoons ground allspice

Cut the meat into strips 12 inches long, 4 inches wide, and 2 inches thick. Remove all membrane from the surface of the meat so that the cure sticks to the moist meat. Mix the curing ingredients and rub the meat on all sides. Then dust on a little more. Thread each strip on a string and hang in a dry, cool place out of the sun. Do not use artificial heat. The venison should hang for a month. After that, it is ready to be eaten even without cooking.

Caribbean Dry Cure

Many Caribbean cure recipes call for a brine flavored with allspice and other ingredients. Although this dry cure is not as spicy as the typical Island brine cure, the salt does the job and the brown sugar adds a distinctive flavor.

3 pounds boneless meat (beef, venison, pork, etc.), in
  1 piece
3 cups sea salt, or more as needed
2 teaspoons brown sugar
1 teaspoon saltpeter

Mix the salt, saltpeter, and brown sugar. Put the meat into a glass container or crock and rub it well with the salt mix. Sprinkle the top with a little more salt, then put in a cool place. After a day, pour off any liquid that has accumulated, turn the meat, and sprinkle on a little more salt mix. Repeat the procedure for a total of 3 days. Then refrigerate the meat for 7 to 10 days.

Before using, freshen this meat by washing it thoroughly and then simmering it for 20 minutes per pound.

Charqui Mendoza

For information on this topic, I am head over heels in debt to *The South American Cook Book*, written half a century ago by Cora, Rose, and Bob Brown. In Argentina, salt-dried red meat is called *charqui*, from which our word *jerky* comes. (To avoid confusion, however, I am considering jerky, covered at the end of this chapter, to be made from thin strips of meat, not from chunks.) In Brazil the word is *xarque*. In the Spanish-speaking West Indies, it is *tasajo*. In French Haiti, it is *boucan*, from which we get the word *buccaneer*.

Of course, there are a number of variations for jerking meat. Here is the technique used in Mendoza, Argentina, where, according to the Browns, not only beef but also guanaco and rhea are made into charqui. (Of course, you can use North American venison or buffalo, or even beef, to make the charqui;

also, U.S. ranchers are now raising emu and ostrich as well as rhea, so these good red meats will be in our markets soon if all goes according to plan. Meanwhile, you might be able to snare a rhea with a bola the next time you go sporting in Argentina.)

Anyhow, to make charqui, select the tender parts of meat and butcher them in large pieces. Slice the pieces into thin steaks and put them under a press to squeeze out excess juice. Rub coarse salt into the meat on both sides. Natural sea salt is best, I think, because it has a good flavor. After salting, stack the pieces on top of each other in a cool place and let stand for 24 hours. Dry the pieces and expose them to the air, covered with a wire screen. After 5 days, put the steaks on a flat surface and pound them with a wooden mallet. Repeat the pounding 2 or 3 times. Then store the charqui in fresh, cool, dry air. Remember that Mendoza is in the Piedmont area of Argentina, not the moist pampas around Buenos Aires. Climate makes a big difference. If you live in New Orleans, you might consider keeping the charqui under the air conditioner's vent.

In South America, charqui isn't merely a trail nibble to be taken along in the backpack. It's a staple for the family, used almost every day in stews of one sort or another, as in the next recipe. Charqui is usually soaked overnight in water to freshen it, then put into the stew pot the following day. In Chile, for example, charquican (a stew made from charqui) contains potatoes, green peas, string beans, yellow squash, chili pepper, onion, tomato, green corn, and charqui. So, make up a batch of charqui and try it in your favorite stew recipe—or in camp chili, made with only charqui, chili powder, tomato paste, and water. It's hard to beat a thick charqui chili served on a plate with black beans, white rice, and chopped onions, mixing it all together as you eat.

In Brazil, xarque is made by first salting the meat in a brine, instead of using a dry cure. It is then dried. Brazilians probably eat a wider variety of cured meats, both salted and smoked, than any other people in the world.

Charqui Stew

A number of South American and West Indian stews call for charqui. Often, the ingredient list is quite long. Here's a representative dish from Argentina, calling for fresh peas or lima beans. In the South, fresh peas or "butter beans" are sold in season and must be shelled; out of season, they can be purchased frozen. Look for frozen baby lima beans, butter beans, butter peas, or field peas, any of which (in my opinion) are better than dried or canned peas or beans.

½ pound charqui (see above)
3 ripe, large tomatoes, sliced
1 medium to large onion, chopped
2 medium potatoes, diced
1 cup fresh peas or lima beans
2 tablespoons white rice
2 tablespoons oil
1 fresh red chili pepper
salt and pepper
¼ teaspoon dried marjoram

Soak the charqui overnight in water. In a stove-top Dutch oven, heat the oil and fry the chili pepper for 10 minutes, stirring well; do not break the pepper open. Remove and discard the pepper. Add the onion, sautéing until soft. Add the tomatoes and simmer for 30 minutes or so. Meanwhile, run the charqui through a food mill or chop it finely. Add the chopped meat to the pot along with 1½ cups water. Add salt, black pepper, and marjoram. Cover and simmer for 1 hour, adding a little more water if needed. Stir in the potatoes, fresh peas or lima beans, and rice. Bring to a boil, reduce heat, cover tightly, and simmer for 30 minutes.

Note: Many South American and West Indian stews call for both jerked beef and salt pork. Usually they also call for cassareep, a condiment made from cassava root and other special ingredients. If you live near Miami or some other market where these ingredients are available, get an authentic Caribbean or South American cookbook and have a look at it. Some of the dishes are traditional, going back to the Arawaks, who cooked them with turtle, armadillo, and other good meats, and to the Caribs, who cooked them with Arawaks and Spanish explorers.

Easy Dry-Cured Brisket

Until I started working on this book, I thought that all corned beef was prepared in a brine. But maybe not. Here's a dry-cure recipe from Morton Salt Company, which they call deli-style corned beef, made with a commercial cure called Morton Tender Quick, which is available in some supermarkets. (If you don't find it in the salt section, look in the canning section.) I can't argue with the results—or with the simplicity of this method. Personally, I prefer to find a fresh brisket that is on the lean side, if possible. The recipe calls for ground bay leaves and ground allspice. If you normally keep these ingredients whole, they can be ground with a mortar and pestle.

1 beef brisket, 4 to 6 pounds
5 tablespoons Morton Tender Quick Mix

2 tablespoons brown sugar

1 tablespoon black pepper

1 teaspoon paprika

1 teaspoon ground bay leaves

1 teaspoon ground allspice

½ teaspoon garlic powder

Trim the brisket and measure its thickness. Mix the cure with the brown sugar, pepper, paprika, bay leaves, allspice, and garlic powder. Rub this mixture into the sides of the brisket, covering the surface. Place the brisket in a plastic bag and tie closed. (I sometimes use large ziplock bags, if the meat will fit nicely.) Refrigerate for 5 days per inch of thickness.

When you are ready to cook, place the brisket in a stove-top Dutch oven and add enough water to cover. Bring to a boil, then reduce the heat and simmer for 3 or 4 hours, or until the meat is very tender.

Note: You can also use this dry-cured brisket with the corned beef recipes in Chapter 4.

Morton's Canadian Bacon

The very lean Canadian bacons are often smoked, but not necessarily so. Here's an easy recipe made with the loin instead of the tenderloin. (If you look at a T-bone pork chop, you'll see 2 rounds of meat. The smaller one, on the bottom, is the tenderloin; the larger one, on top, is the loin.) You can often purchase a whole pork loin at the supermarket, or talk to a local meat processor. It should be very fresh and properly chilled.

This recipe is designed to make use of such conveniences as the plastic bag, the modern refrigerator, and Morton Tender Quick Mix, a commercial cure.

1 boneless fresh pork loin
1 tablespoon Morton Tender Quick Mix per pound of pork
1 teaspoon sugar per pound of pork

Thoroughly mix the cure and the sugar. Trim the pork loin. Rub the cure and sugar mixture into the meat, covering all exposed areas. Put the loin in a plastic bag, tie it shut, and put it into the refrigerator for 3 to 5 days. Remove the loin from the bag and soak it in cool water for 30 minutes. Pat the loin dry, then refrigerate it uncovered to allow it to dry slightly before cooking.

To cook the Canadian bacon, slice it into ⅛-inch slices and fry it in a little oil over low heat, turning to brown both sides. The cooking should take 8 to 10 minutes.

Variations: Use brown sugar instead of regular sugar. After the cure, remove the loin from the bag and smoke it for several hours.

If you want "pea meal" bacon, dry the loin as directed and rub it with a mixture of stone-ground cornmeal, black pepper, and cayenne. Cover the loin with plastic wrap and refrigerate. When you are ready to cook the bacon, slice it ⅛ inch thick, sprinkle with fine stone-ground cornmeal, and fry as directed.

JERKY AND PEMMICAN

Note that the jerked chunks of meat in the previous recipes are cured with lots of salt. For the most part, Native Americans didn't use salt in their jerky and usually didn't use large chunks of meat. Instead, they cut the meat into thin strips and dried it in the air and sun. Excellent jerky can still be made with this no-salt technique. Dry climate is a key to success.

Almost always, it is best to cut the meat into strips about ⅜ or ¼ inch thick. If the jerky is to be eaten as is or cooked in stews, cut the strips with the grain of the meat. If it is to be pulverized for pemmican or some other recipe, it's best to cut the strips across the grain. After drying, the jerky can be stored in jars or plastic containers. It should be kept in a dry, cool, dark place until needed. It doesn't have to be refrigerated.

The comments and recipes in this chapter reflect the old ways. Modern man has figured out techniques to speed up the process, first by hanging the jerky over a fire, then by putting it into an oven on low heat for half a day, or in a patio electrically heated "smoker," and finally by zapping it in a microwave oven. A thousand recipes for modern ways can be found in other books and magazine articles, calling for all manner of stuff such as Liquid Smoke and Worcestershire sauce. In this book I have tried to stick to the basics of both technique and ingredients.

Modern Americans view jerky as noshing fare or as a trail nibble. More often than not, however, Native Americans and early settlers used it as the main meal in stews and soups, or pulverized and mixed it with dried fruits and grease. Some native peoples even made baby food by mixing pulverized jerky with crushed butternuts and hickory nuts; this mixture was dried and, when needed, mixed with boiling water and fed warm to infants. One historically famous mixture—pemmican—is covered at the end of this chapter.

Indian Jerky

If you have a dry climate and want to try Indian jerky, cut the meat with the grain into thin strips. Stretch (or jerk) each strip and hang it over a line, or run a length of cotton twine through a hole in one end and fix the twine to a line. Leave the jerky in the sun all day, but take it inside at night and rehang it the next morning. The jerky will be ready in 3 or 4 days. You can also make this jerky in an enclosed area, such as a porch, if you have dry air with good circulation. A screened porch with morning or afternoon sun is ideal, especially if you have a breeze or a fan. The screen also helps keep the insects off the meat.

Although this method works under good conditions, I recommend that the strips of meat be soaked overnight in brine (1 cup salt to 1 gallon water) before hanging them. The salt will add flavor as well as help cure the meat. If you want to cheat a little further, hang the meat in your oven, turn the heat to the lowest setting, and leave the door ajar until the meat is dry. How dry? If you are going to use the jerky in soups and stews (or in pemmican), it can be very dry, even brittle. For chewing, however, the strips should be dry but still flexible.

Wyoming Elk Jerky

This recipe goes back for at least two generations to somebody by the name of Payne, at Green River, Wyoming. About 4 or 5 pounds of elk (or other good red meat) is cut into strips. The strips are put down with plenty of salt between each layer. Then 1 cup brown sugar is dissolved in water, along with 1 tablespoon saltpeter, 1 tablespoon ground allspice, and ½ teaspoon red pepper. This solution is poured over the salted meat, and the meat is left in the cure for 36 hours.

Using a darning needle with cotton twine, a loop is made on each strip. Then the strips are strung on a line, with the aid of clothespins, and left to dry in the sun. If you prefer, you can put the meat in your oven, turn the heat to the lowest setting, keep the door ajar, and leave it overnight, or until the jerky is dry.

Pemmican

Anyone who doesn't have all of his teeth, or who is merely tired of chewing on jerky, ought to consider making a batch or two of pemmican. The term came from the Cree language and means journey meat. The idea behind pemmican, of course, is to pack a maximum amount of nourishment into a minimum bulk.

To make pemmican, the Crees and other Native Americans hung thin strips of venison or buffalo in the sun for a few days, as when making jerky. After the meat dried, they pounded it into a pulp and mixed it with the fat from a bear or a goose. (Modern practitioners might use hog lard.) Such a mix, packed in rawhide bags, would keep indefinitely. Sometimes dried fruits or berries were mixed in with the meat and fat, and became part of the pemmican. Or the fruits could be stored separately and eaten along with the pemmican.

Note that lean beef can be used for making pemmican if it is properly dried. Some Native Americans even used dried fish for making pemmican. Also, remember that the dried berries can include such wild species as buffaloberry. In many areas, ground nuts were added to the mix. Any recipe for beef or venison jerky will do. In other words, merely pound or grind the jerky and stir it in with the fat. Roughly, the mix is 1 part fat to 2 parts dried meat.

Learning from the native peoples, the early explorers also made good use of pemmican. The French explorers used it extensively, and the term even gained a brief entry in *Larousse Gastronomique,* which specified that the rump of the animal be used. After the French got hold of the recipe, all manner of spices and such were added. Even cinnamon.

Pemmican

Modern backpackers can make up some pemmican at home, then take it with them on a journey. Of course, the meat can be dried in the kitchen oven as well as on the clothesline. These days most of us don't have rawhide bags and would prefer to keep the pemmican in what we perceive to be a more sanitary manner, as set forth in the following directions:

1 pound dried meat or jerky
½ pound lard or other fat that doesn't require refrigeration
½ pound dried apricots or similar dried fruit or berries
salt and pepper to taste
melted paraffin
cheesecloth

Pulverize the dried meat and fruit; I usually use a mortar and pestle for this, but a food grinder or a food processor will help for the initial work. Mix the pulverized meat and fruit into the lard, adding a little salt and pepper. Form the thick paste into small bars about 1 inch in diameter and 3 to 4 inches long. Wrap each bar in cheesecloth and dip it quickly into melted paraffin.

It is, of course, always best to store pemmican in a cool, dry place, but refrigeration isn't required. When you're taking pemmican on a journey, pack it in a cool spot that isn't exposed to the sun. Native Americans and the early explorers ate pemmican raw, like a candy bar, and sometimes they boiled it in water, making a gruel. It can also be mixed into soups and stews, along with such ingredients as squash blooms and strips of dried pumpkin. Either way, pemmican is a highly nutritious mix for camp or trail. And for home consumption, too.

Carne Adobado

To the native peoples of the Southwest and Mexico, the chili pepper was a very important part of the diet and was often used in recipes for stews and soups—as, for example, in chili con carne. When the seeds and core are removed, peppers provide flavor and sustenance without too much heat. Usually, dried red peppers are boiled for 1 hour, then mashed into a paste or puree. For a mild puree, the seeds and core are removed; for a hot puree, they are left in. I recommend the hot kind for this recipe. As a rule, it takes about 24 dried chili peppers to make enough sauce for the recipe, but, of course, a good deal depends on the size and strength of the peppers. If you don't raise your own peppers, you can buy a string of them (ristra) at some supermarkets and by mail order. The recipe calls for oregano, which was brought to the Southwest by Europeans. If you are a stickler for authenticity, leave it out.

   3 pounds fresh lean pork
   2 cups red chili pepper puree (see above)
   1 clove garlic, mashed
   2 teaspoons salt
   1 tablespoon dried oregano

Cut the pork into strips about 6 inches long and 2 inches square. Mix the remaining ingredients and toss with the pork to cover all sides of the meat. Marinate in a cool place for 24 hours. When ready, hang the pork strips in a cool place to dry. When thoroughly dry, store the meat in a cool place.

   To cook the dried meat, pound it with a meat mallet or the back of a cleaver until well shredded. Add some fresh chili sauce and cook until tender.

4

CORNED BEEF & OTHER BRINE-PICKLED MEATS

One of my favorite recipes for corned beef, and for corned venison, comes from George Leonard Herter's rather outlandish *Bull Cook and Authentic Historical Recipes and Practices.* Herter seems to think that the name *corned beef* stems from a mistaken link to corn whiskey. Many Americans also tentatively associate the term with corn whiskey, usually without knowing why. By way of explanation, Herter says that corned beef originated in London in 1725. During World War II, he goes on, South American beef was shipped to the U.S. fighting forces in Europe. The troops gave it the name *corned Willie,* meaning goat meat cured by soaking it in corn whiskey. But, Herter says, corn whiskey was not and is not used for corning beef. He's right about the corn whiskey, and maybe about London, but the rest of his derivation is questionable.

I think the confusion comes from the different meanings of the word *corn.* In England, the word originally meant grain, which would include wheat and oats and barley. The term was in widespread use long before the discovery of American corn, or maize. In fact, the term was so common that it was also used to denote anything in small pieces, just as we now say a grain of sand. The salt used in early times was usually in the form of grains, much like rock salt today, and meat cured with the aid of these grains of salt was said to be corned. So there you have it.

Before proceeding, however, I must point out that water is a very important ingredient in corned meats. It doesn't have to be distilled, but it should have a clean taste. I've seen water in towns that has too much stuff added to it, whereas spring water in the area was excellent. Also, in parts of Florida and no doubt some other areas, the water tastes of sulfur. Soaking a piece of meat in this stuff doesn't help the flavor.

Corned Beef According to Herter

Here's the "authentic historical" recipe for corned beef, which Herter claims to have published for the first time. To make the full measure of corning liquid, you'll need a container that will hold 6 gallons. It's best to pour in 6 gallons of water, then mark the level. In Herter's recipe, you'll end up with 6 gallons of corning liquid, including the water and other ingredients. This liquid is used to completely cover the meat. Reduce the measures if you don't need 6 gallons. Before dismissing the 6-gallon figure, however, remember that it's easy to corn venison as well as beef. Corning is an excellent way to solve a temporary storage problem with a fresh-killed elk or moose, and, of course, large chunks of beef can be purchased from meat processors. Beef briskets are popular for corning, but better and leaner cuts of beef can be used to advantage. I like sirloin tips cut into 4- to 6-pound chunks.

The container for the pickling liquid must be nonmetallic. Try a pickling crock or an old churn. Even a plastic or Styrofoam ice chest will work.

1 4- to 8-pound chunk of beef
3 pounds salt
1 large onion, minced

4 cloves garlic, minced
1½ cups sugar
4 tablespoons mixed pickling spice
1 tablespoon black pepper
1 teaspoon ground cloves
6 bay leaves
2 ounces sodium nitrate
½ ounce sodium nitrite

Place all the ingredients except the meat into the crock. Add enough water to make a total of 6 gallons. Add the meat, making sure that it is completely covered. Place a plate or wooden plank over the meat and if necessary weight it with a stone (or something nonmetallic) to keep the meat submerged. Corn the meat for a total of 15 days. On the 5th and 10th days, remove the meat, stir the corning liquid with a wooden paddle, and repack the meat. After the 15th day, remove the meat. Use some of the meat immediately, if needed, and store the rest in the refrigerator or freezer.

Ideally, the meat should be corned at a temperature of about 38°F. It can be corned at a higher temperature, but, Herter says, more salt should be added. For every 15-degree increase in temperature, increase the salt measure by one third.

To cook the corned beef, cover it with water, bring it to a boil, skim the surface, reduce the heat, and simmer for 5 hours or so, until the meat is fork-tender. Or use the meat in one of the recipes in the next section.

A. D.'s Corned Meat

I have a tendency to go back to the basics on many traditional dishes. Here's my authentic historical recipe. Admittedly, sea salt is on the expensive side these days, but remember that spices and other additives are eliminated, so that the recipe is comparatively inexpensive. More important, sea salt gives the meat a distinctive flavor.

1 5-pound chunk lean beef or venison
3 cups sea salt
1 gallon water

Wash the meat and fit it into a nonmetallic container just large enough to hold it, but not too tightly. Mix the sea salt into the water, then pour the brine over the meat. Place a plate or saucer over the meat, then weight it down with a jar of water so that the meat will be completely submerged. Cover the container with a cloth and put it in a cool place or into the refrigerator for 10 days.

Rinse the corned meat and boil it in fresh water (without spices) until it is tender. Use the meat as a roast or slice it for sandwiches.

Caribbean Pickle

This salt-pickle recipe has been adapted from *The Complete Book of Caribbean Cooking* by Elizabeth Lambert Ortiz. Note that the brine contains allspice, a native West Indian ingredient. The popular jerked meats of Jamaica also contain lots of allspice, but these depend more on pepper than on salt.

This recipe can be used for either beef, pork, or venison. Or mutton, for that matter. The measures are just right for 5 pounds of meat, and I suggest that you try a lean roast for making this Caribbean version of corned beef.

4 to 5 pounds fresh lean meat, in 1 piece
1 gallon water

3 cups sea salt
2 teaspoons saltpeter
½ cup dark brown sugar (from the Islands, if available)
4-inch stick cinnamon
1 tablespoon allspice berries
1 tablespoon powdered mustard
1 medium onion, sliced
1 tablespoon minced fresh thyme
1 or 2 hot red chili peppers

Put the water into a large saucepan. Mix in the mustard, all-spice, cinnamon, brown sugar, salt, and saltpeter. Boil for a few minutes, then let cool and skim the surface. Put the meat into a stoneware crock or other nonmetallic container. (Try a large crockpot.) Pour the pickling solution over the meat, then add the thyme, onion slices, and chili peppers. Place a plate, saucer, or block of wood over the meat so that it is completely submerged in the brine. Leave the crock in a cool place for 10 days, turn-ing the meat once a day. Remove the meat from the pickle and refrigerate it until needed.

Simmer the meat in water for several hours, or until it is very tender. Use the meat as a roast or slice it for sandwiches.

Old Virginia Corned Beef

Here's an old beef cure that I have adapted from Mary Ran-dolph's *The Virginia Housewife*. Although the recipe is primarily of historical interest, the method can be used for corning large batches of beef or large game such as elk or moose.

Prepare your brine in the middle of October, the book says. First, get a 30-gallon cask, take out 1 head, drive in the bung, and put some pitch on it to prevent leaking. See that the cask is tight and clean. Put into it 1 pound of powdered saltpeter, 15 quarts of salt, and 15 gallons of cold water, stirring as you go. Stir until the salt is dissolved. Put a thick cloth over the cask to keep out the dust. (These proportions have been accurately ascertained—15 gallons of cold water will exactly hold, in solution, 15 quarts of good, clean Liverpool salt and 1 pound of saltpeter; this brine will be strong enough to float an egg.) The brine will cure all the beef that a private family can use in the course of the winter, and requires nothing more to be done except occasionally skimming the dross that rises. It must be kept in a cool, dry place. For salting your beef, get a molasses hogshead and saw it in half, so that the beef may have space to line on; bore some holes in the bottom of these tubs and raise them on one side about 1 inch, so that the bloody brine may run off.

Be sure that your beef is newly killed. Rub each piece very well with good Liverpool salt. A vast deal depends upon rubbing the salt into every part. It is unnecessary to put saltpeter on the beef. Sprinkle a good deal of salt on the bottom of the tub. When the beef is well salted, lay it in the tub, and be sure you put the fleshy side downward. Put a great deal of salt on your beef after it is packed in the tub; this protects it from animals who might eat it, if they could smell it, and does not waste the salt, for the beef can dissolve a certain portion. You must let the beef lie in salt for 10 days, then take it out, brush off the salt, and wipe it with a damp cloth; put it in the brine with a bit of board and weight to keep it under. In about 10 days it will look red and be fit for the table, but it will be red much sooner when the brine becomes older. The best time to begin to salt beef is the latter end of October, if the weather be cool, and from that time have it in succession. When your beef is taken out of the tub, stir the salt about to dry, that it may be ready for the next pieces. Tongues are cured in the same manner.

RECIPES FOR CORNED BEEF

Corned beef is quite rich, so that a quarter pound will be a sufficient serving for a person of normal appetite. Most of these recipes call for cooking a 4- to 6-pound chunk of meat, which is more than enough for the average family of today. Leftovers can be sliced for sandwiches or used in hash.

Corned Beef and Cabbage

I don't know where this dish originated, but Ireland is my guess. In any case, the combination of corned beef and cabbage is a happy one, at least to my taste.

    4 or 5 pounds corned beef brisket
    1 large head green cabbage
    10 or 12 small onions, about 1 inch in diameter
    1 tablespoon chopped fresh parsley
    1 teaspoon powdered mustard
    2 bay leaves
    10 black peppercorns
    4 whole cloves

Wash the corned beef and put it into a Dutch oven or other container suitable for long, slow cooking. Cover the meat with water, then add the parsley, mustard, bay leaves, peppercorns, and cloves. Bring to a quick boil, reduce the heat, cover, and simmer for 3 hours. Skim off any fat that has come to the top, and pour off some of the water, so that the corned beef is only half covered. (The idea is to steam the cabbage instead of boiling it.) Peel the onions and put them around the sides of the meat. Wash and cut the cabbage into wedges, allowing at least 1 piece for each person, and put them on top of the meat. Cover tightly and simmer for 30 minutes, or until the cabbage is tender.

Leftovers can be refrigerated and heated later. If I have plenty of good bread, preferably corn pone, I can make a complete meal of leftover corned beef and cabbage. The stock from the corned beef and cabbage can be saved and used in soups and stews.

New England Boiled Dinner

This is a pretty dish as well as a tasty one, and I like to serve it on a family platter about 20 inches long. The cooked corned beef is placed in the center with the other vegetables arranged all around.

    4 pounds corned beef or venison
    4 ounces salt pork, cubed
    6 to 8 medium onions, golf-ball size
    6 medium potatoes
    2 medium turnip roots
    6 carrots, scraped
    2 small heads green cabbage, quartered
    3 or 4 beets, cooked separately and sliced
    1 tablespoon chopped fresh basil

1 teaspoon freshly ground black pepper
2 bay leaves

Rinse the corned meat, put it in a pot along with the salt pork, cover with water, and boil for 10 minutes. Pour off the water, cover again with cold water, and bring to a new boil. Add the bay leaves, onions, pepper, and basil. Cover tightly and simmer (but do not boil) for 5 or 6 hours, or until fork-tender. Add the potatoes, carrots, and turnips; cover and cook until the potatoes are done. Add the cabbage, cover tightly, and simmer for 10 minutes. Note that it is not necessary to cover the cabbage with liquid; if you cover the pot tightly, the steam will cook it.

Center the corned beef on a large platter and arrange the vegetables around it in groups. Serve hot with corn bread.

Note: There are many variations in vegetables and mixes. Try rutabagas, peeled and cubed, instead of turnips. Also try parsnips. A corned beef dish of old Virginia called for pumpkin cut into wedges. I have cooked the dish with corn on the cob and with Jerusalem artichokes. Also, Brussels sprouts can be used if you cut back on the cabbage.

Boiled Dinner, the Maine Way

People in Maine have firm opinions and often back them up. Once, for example, the Maine legislature passed a law prohibiting the use of tomatoes in anything called clam chowder. Maine people also have opinions about corned beef. A book called *Good Maine Food* says that many families on the Maine coast corn their own beef, and "according to those Maine cooks whose reputations have gone farthest," it is best made from the thick rib or from a piece of flank next to the loin instead of from the brisket. The meat is salted down, then immersed in a brine. Some people pickle the meat for as long as a month; others pickle it overnight.

After pickling, the meat is boiled until tender, about 2 hours. Then the meat is removed from the brine, drained, and chilled. It will be served cold. The brine is saved until time to cook the dinner. To prepare dinners, start a countdown for a 2-hour period. Bring the liquid from the corned beef pot to a boil. After ½ hour, scrape 2 carrots, cut them into pieces, and add them to the pot. After 1 hour, add 1 sliced turnip and 1 small green cabbage cut into as many pieces as there are persons to be served. After 1½ hours add 8 peeled potatoes, halved, and a sliced summer squash. At the end of 2 hours, remove all the vegetables. Put the cold corned beef on a platter and arrange the vegetables all around. Add 2 cups of sliced beets, which have been cooked separately so that they won't discolor the other vegetables. Serve with tomato catsup or vinegar.

Maine Corned Beef Hash

For this fine dish I also owe *Good Maine Food*. The recipe is preceded by a dialogue from a Mrs. Ivy Gandy to a Mr. Roberts:

"But how, my dear Mr. Roberts, can we have corned beef hash unless we first have real Maine corned beef? Didn't you know that nowhere in all the world do they know how to corn beef as it is done in Maine? All up and down the length of Long Island I have sought a meat market that has real corned beef, but all I can find is a tough length of reddened fibres, salty and bitter. My corned beef hash is made from meat that I corn myself by the old Maine recipe."

The hash is best when cooked in a large cast-iron skillet.

4 cups chopped cooked corn beef
6 medium potatoes
¼ cup butter
½ cup boiling water
salt and pepper to taste
catsup

Boil the potatoes in their skins until tender. Cool, peel, and dice. In a large wooden bowl, mix the diced potatoes with the corned beef and chop everything again until very fine. Melt the butter in a large skillet, then add the boiling water and corned beef mixture, stirring as you go. Add a little salt and pepper. Cook on low heat until a crust forms on the bottom. Fold the mixture over like an omelette. Serve hot with catsup.

Brine-Cured Pork Hams and Shoulders

Although large hams can be cured in a pickle instead of being dry-cured, the process really works best for the small hams and shoulders. One advantage of a pickle over a dry cure is that it allows the use of liquids such as honey and wine. Wine is used in some Italian cures and in the Spanish Serrano hams. And the Suffolk hams of England are pickled in stout.

One problem with curing large hams, and large numbers of hams, in a pickle is that salt penetration is slow and the liquid drawn out of the ham tends to dilute the pickle. Large hams are best dry-cured at first, then transferred to a pickle. Shoulders and small hams (under 10 pounds) can be put directly into the pickle, which is what I recommend. (In other words, with small hams and shoulders you can omit steps 2 and 4 below.)

My favorite recipe is of Italian origin, but for it I am indebted to *The Country Kitchen,* a British cookbook by Jocasta Innes. The steps below can be completed with ordinary kitchen equipment, making this recipe ideal for curing a single shoulder. The recipe can also be used for large batches of pork, provided the pickle is kept up to strength. In any case, this is a good recipe for beginners as well as gourmets. Ms. Innes seems to be fond of scalding things, and I have left these instructions in the procedure, partly to appease my mother-in-law, a kindred spirit who scalds at every opportunity and who is still mad at me over my comments in my *Cast-Iron Cooking* book about her scalding and scrubbing my favorite skillet, thereby ruining the seasoning.

1. Rub sea salt into the surface of the ham, using 1½ pounds of salt for a 12-pound piece. Mash the salt into the cavity around the bone on either end.

2. Scald a nonmetallic container of suitable size. Scald also a wooden lid, or plank of suitable size, and scald a stone for weight.

3. Put the ham into the container, place the plank on it, and weight the plank with the stone. Leave the container in a cool, clean place overnight. (If you can make room, use the refrigerator, as I do.) If you have a large ham, say 18 or 20 pounds or so, salt it down for an extra day or two, rubbing in more salt after the first day.

4. Mix a pickle by boiling 4 pints of good water and adding 2 pounds salt, 1 tablespoon baking soda, and

2 tablespoons saltpeter or sodium nitrate. Turn off the heat and add 4 pints white wine, 1 tablespoon allspice, and ½ tablespoon black peppercorns. Set aside to cool.

5. Remove the ham from the container and drain off any liquid drawn out by the salt cure. Wipe the ham and put it back into the container. It's best to have the ham slightly cooler than the pickle. (If the pickle were colder, Ms. Innes says, it would continue the extracting process rather than the permeating one.) Pour the pickle over the ham, straining it through a muslin bag. The pickle should completely cover the ham. If not, use a smaller container or make more pickle.

6. Place the wooden lid on the ham and weight it down with the stone. Leave it in a cool place for a total of 28 days.

7. After each 7 days, test the pickle for strength. Do this by putting a fresh egg into the solution. If it half sinks, the pickle needs more salt. Boil about ⅓ the total amount of fresh water with 3 times as much salt as the original pickle. Cool this mixture before adding it to the container.

8. Every 4 or 5 days, remove the ham and scald a wooden spoon. Stir the brine with the wooden spoon and replace the ham. Scald the wooden lid and scald the stone, then replace both.

9. Remove the ham after 28 days and scrub off the surface salt with a brush.

10. Hang the ham to dry in a cool, airy place for 7 days.

11. To cook the ham, put it in a pot of plain water, bring it to a boil, reduce the heat to very low, and simmer it for 2 hours. Then simmer it for 1 hour in clean water, along with chopped celery, carrots, onions, and so on for flavor.

Note: I highly recommend that you try this procedure with 2 shoulders, or with a single shoulder divided into a picnic and a butt cut. (Also, you can cut a whole ham into a butt and a shank portion.) By having more than one piece of meat, you can cook one as soon as it is ready and hang one for a month or so.

I might add that similar brine-cured hams are eaten in other parts of the world, often at Christmas dinner. (In other countries, turkey isn't as important as it is in America for Christmas and Thanksgiving.) In Finland and Sweden, the *grilijerad shinke* is put in brine weeks ahead of Christmas. It is served hot on Christmas Eve and cold on the following days. Finns serve baked rutabaga or turnip roots with the ham, whereas Swedes prefer to serve the ham with boiled red cabbage and mashed potatoes.

Sweet Pickle for Pork

Here's an old Wyoming recipe. The measures are for 100 pounds of pork, but they can be scaled down for smaller amounts. Before applying the pickle, the meat is rubbed with dry salt and packed in wooden barrels or wooden boxes.

   7 gallons water
   salt
   3 pounds sugar
   2 ounces saltpeter
   1 ounce ground red pepper

Put the water in a large pot and dissolve salt in it until it will float an egg. Stir in the remaining ingredients and bring to a boil. Let the brine cool, then pour it over the salted pork. Let stand for 10 days. Drain off the brine, bring it to a boil, let it cool, and pour it over the pork again. Cook the pork as needed.

PART TWO

Cold-Smoking Meat, Fish, & Game

I GREW UP ON A FARM IN EAST TENNESSEE WHERE WE
KILLED HOGS EVERY FALL. THE HAMS WERE SUGAR-
CURED AND SMOKED WITH HICKORY AND SASSAFRAS.
I SURE WISH I COULD GET JUST ONE OF THOSE HAMS!

—*Dr. Eph Wilkinson*

5

SMOKEHOUSES & RIGS

On the family farm where I was raised, we had a smokehouse of the old-timey sort. My grandfather, Jeff Livingston, built it from rough, unfinished lumber hewn from the tall pines that he had cleared from the land. During my lifetime the smokehouse has had a tin roof put on, with lead-capped nails, but I am certain that the original roof, like that of the main house, was made of wooden shingles. Believe it or not, Grandpa put the family tub in the smokehouse. The main house didn't have a bathroom because there was no electricity or running water. The smokehouse was beside the well, making it more convenient for the tub. Later, when an electric pump and running water became available, the tub was moved into the main house.

The fire for the smokehouse was built in a shallow trench in the dirt floor. First, we built a hot fire with cured oak or hickory, burning it down to red-hot coals. These coals were concentrated in the trench; then two green hickory logs about 6 or 8 inches in diameter, as I remember them, were laid on top of the coals. Of course, the logs were the length of the trench. I emphasize that the limbs were green, freshly cut especially for smoking. Gradually, the green limbs would themselves become coals and would have to be replaced by new green limbs to provide smoke.

The idea, when laying such a fire, is to produce lots of smoke without having to constantly add more wood. The fire should not blaze up and burn, producing too much heat for cold-smoking and using up too much wood, nor go completely out. Much depended on how close the green limbs were to each other and on the sloping sides of the trench. My father was an expert at laying such a fire. He was, it seemed to me, quite fussy about the details—and with good reason. On the one hand, he didn't like to go outside in freezing weather after 9 o'clock at night to tend the fire. On the other hand, he wanted to see smoke when the new day dawned. Whether or not we had smoke all the time was a matter of great importance to us back then, although in more recent years I have decided that letting the smoke expire for several hours won't hurt a thing in cold weather. Just don't tell Pa.

Our smokehouse was about 12 feet wide and 14 feet long. I doubt that Grandpa followed a set of plans or exact dimensions. He just built it. (He built the fourteen-room main house the same way, framing it on 14-inch sills, and there wasn't a square corner in it, as I found out when trying to remodel it.) The sides of the smokehouse were made of 12- or 14-inch rough-hewn pine boards, nailed to the frame with square nails. The boards had a crack between them. This crack wasn't there because Grandpa didn't fit the boards properly before nailing them; they were spaced by design so that the smoke could get out. The smokehouse had exposed rafters, and the meat was hung from cross timbers well above a strapping boy's head. I don't remember Grandpa's technique, but to tie the hams, Pa used thin strips (withes) of green hickory or white oak. My wife tells me that her father, on a similar farm in a contiguous county, used strips from the bladelike leaves of the yucca, which they called bear grass.

Another very important feature of our smokehouse was a wooden workbench built into one side. Just as the walls contained cracks for smoke to escape, the worktable contained cracks between the boards to drain the salty juices that seep out of the curing meat. (In time, the dirt inside the smokehouse became quite salty and would quickly rust hoe blades and axes if they were left there.) After being properly salted, the meat was grouped on the boards according to size and kind, then left on the boards to cure.

A similar smokehouse can be built today, and I have seen them converted from existing structures. (I know one fellow who inherited the family farm and converted an old two-seater outhouse for smoking. He said it worked pretty well, but somehow the sausages didn't come out quite as good as planned. He allowed that part of the problem might have been in his head.) Most modern families, rigged with mechanical refrigerators and freezers and supermarkets, won't require a large log-burning smokehouse and might therefore consider making a smaller structure. Instead of burning logs, you can heat chunks of hickory on electric hot plates or perhaps on a couple of charcoal-burning hibachi grills.

In any case, I am setting forth several suggestions for smokehouses. Some of these are obviously temporary, but some are compatible with modern houses and even suburban landscapes. I suggest that the beginner start with an inexpensive, temporary structure and use it until he gains some smokehouse experience. Then he can customize one of my ideas, or, better, design a smokehouse from scratch. Note that other uses of a smokehouse are possible. The modern practitioner probably won't have to put the family bathtub in the smokehouse, but the deer hunter or the farmer who butchers a goat might want to consider the structure as a place to hang a whitetail or goat for aging. In this case, it would be very important to cover all vents and openings with screening to keep the flies away.

The sketches in this chapter are merely suggestions, not detailed plans. I assume that any local builder or competent do-it-yourselfer can provide the construction details to suit local building codes, individual preferences, and landscape compatibility as well as available materials and funds. It is much more important to consider the function of a smokehouse and what it must do to accomplish cold-smoking—and what it must *not* do.

1. The smokehouse must *not* be an oven. That is, it must not get hot enough to cook the fish or meat in its normal operating environment. For cold-smoking, the smokehouse should remain below 100°F and preferably between 70°F and 80°F. This requirement of enough heat to produce smoke but not enough to raise the temperature to cooking levels can be accomplished in several ways, usually involving either a large smokehouse with a small fire or some scheme to generate the smoke outside the smokehouse and pipe it in.

2. The smokehouse must be a smoke chamber. That is, it must hold the smoke densely. On the other hand, the smoke must escape somewhere. It's best to have the exhaust openings at the apex, or in the top of the structure, or near the top, so that there is a slow flow of smoke from bottom to top. The fish or meats to be smoked should, of course, be hung or racked in the flow of smoke.

3. The smokehouse must have some means of holding the meat, either on racks or suspended

from poles or rafters. Obviously, this requirement will be of great importance to those who smoke meat commercially, in which case the smokehouse should be designed to accommodate the maximum amount of meat.

4. For cold-smoking, the smokehouse must function safely for extended periods. of time.

A. D.'s Walk-in Smoker

My favorite design is somewhat different from Grandpa's. The drawings here show the shape better than I can describe it. Note that the low side has a wooden workbench (best with cracks between the boards for drainage). The high side has a pole (or poles) in the top for hanging hams and whole fish. It can also be fitted with racks, but access is not as convenient.

This smokehouse can be made as long or as short as you wish, but I recommend *at least* 8 feet for smoking so that you'll have room to move about. If you want a storage shed, consider making the structure 16 feet or longer and then dividing it in half, with a door on either end. The structure can be framed with 2-by-4s. On the roof, wooden shingles would be great. I also like the sides made with overlapping wooden boards. All of the wood can be left unpainted, if that décor jibes with your landscape. The inside walls of a smokehouse are usually bare, and the rafters are exposed.

Modern conveniences can include electric wiring. Also, vents can be built in at ground level using standard units made for venting boxed-in houses. These usually have screening. Similar

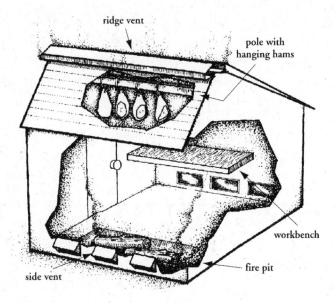

A walk-in smoker

vents can be placed around the top of the smokehouse, under the roof overhang, but I really prefer to use a ridge vent along the top. (Ridge vents are available at building supply houses. They are a sort of vented cap that is placed over the apex of the roof, as shown in the sketch.)

If you prefer to make such a structure with a standard roof, consider an overhang design. This can be used for a woodshed or for storing garden tools.

If you plan to use wood or charcoal for fuel, consider putting the structure on the ground, with a dirt floor. A concrete footing can be used for a foundation.

CHIMNEY SMOKING

According to my copy of *Larousse Gastronomique,* cold-smoking is (or was) used almost exclusively in France for herring, which are first pickled in salt for a certain length of time, depending on how they are to be used. Bloaters, for example, are only lightly salted and will keep for only a few days. Pickled smoked herring, on the other hand, are salted for at least 8 days. After pickling, the herring are strung up on rods and placed in a chimney. A fire of beech wood is lit, which causes a rush of hot air up the chimney. This draught dries the herring quickly. Then the fire is dampened with beech sawdust and chips, which creates a dense smoke to flavor and color the herring.

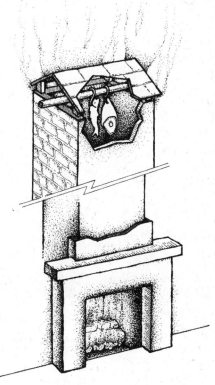

Chimney smoking

I've never seen a commercial smoking chimney, but reading the French text gave me some ideas for home application. As it happens, I have a fireplace in my den and the large chimney emerges on a rather flat roof with easy access. In short, I have hung fish on rods and suspended them into my chimney from the top. Usually, I first build a fire with oak or pecan wood and let it burn down before I load the fish. Then I dampen the fire with green wood. For true and safe cold-smoking, the temperature of the air that exits the chimney during smoking should not exceed 100°F. It's easy enough to hang a thermometer and check it.

In hot weather, you may have to forget the fire, turn on the air conditioner, and put a two-burner electric hot place into the fireplace. Put a cast-iron skillet on each burner and fill it with green wood chips or partly fill it with hardwood sawdust that has been soaked in water. Adjust the heat on the hot plate to make smoke without burning the wood.

Try this method with small fish of about 1 pound. Larger fish should be filleted and smoked on a rack instead of being hung from a rod. In any case, be certain that your fish won't drop into the chimney.

Note that you can also smoke sausages and other meats by this method. But don't allow too much grease to drip into your chimney. If you do, you may be setting the stage for a roaring and dangerous fire.

BACKYARD BARBECUE SMOKERS

Some time ago, agricultural engineers with the U.S. Department of Agriculture published some drawings of various smokehouses, most of which look quite like a typical country outhouse. Some of these plans, such as the one depicted here, were then reprinted in a book called *Butchering, Processing and Preservation of Meat,* by Frank G. Ashbrook. While all of the schemes will no

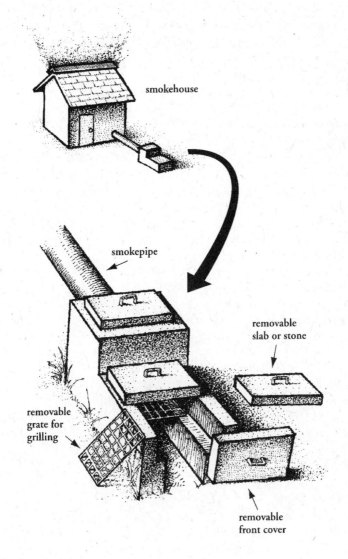

smokehouse

smokepipe

removable
slab or stone

removable
grate for
grilling

removable
front cover

A backyard smoker

doubt work, I am especially attracted to a design for combining a smokehouse and an outdoor grill. This could be a practical unit for the backyard, especially if old bricks or stones are used in the construction. The details here show the firebox and grill, with a flue leading back to the smokehouse.

TEPEE AND WIGWAM SMOKERS

The more or less conical shape of the tepee used by the nomadic Plains Indians has a hole in the top to allow smoke to escape. This same shape can be used for smoking meat and fish, and is

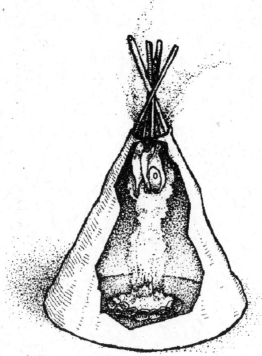

A tepee smoker

a good design for smoking a single ham or a few fish. It is not a good design for smoking a large batch of fish or meat, however. A wigwam structure popular with the Algonquian people can also be used.

With a tepee, three main poles are joined at the top and spread at the bottom like a tripod. Then more poles are spaced to round out the structure. Then the poles are covered with hides or some other material, leaving a hole at the top. (A real tepee doesn't have the hole at the very top; rather, it is off center, allowing for a flap to cover the hole.) A smoke fire is built on the ground, and the meat is suspended from the top. The covering for a temporary tepee can be made from canvas. Even plastic sheeting can be used if the tepee is large and the fire small and confined to the center.

Note that a tepee is a portable, breakdown structure. The wigwam, on the other hand, is built with poles inserted into holes in the ground, or from sapling trees that can be bent over. This produces more of a dome shape, but it can still be used for smoking, simply by hanging the meat or fish near the vent hole in the top.

A-FRAME SMOKERS

An A-frame structure is relatively easy to make and can be permanent, temporary, or collapsible and portable. In the latter construction, the triangular ends would be removable and the two sides hinged at the top; thus, it can be partly disassembled and folded for storage. It must, of course, have a vent or vents at the top, which is easy to accomplish at the apex where the two sides join together. If the structure is permanent, it should be topped with a ridge vent. You can make a walk-in smoker 8 feet long with five sheets of plywood, and a few 2-by-4s. For a permanent structure, of course, the outside plywood should be covered with shingles or some weatherproof material. For a really nice structure, consider using natural wood siding instead of plywood. Note that you can build a worktable and shelves at one end of the structure.

The fire is built on the ground. If you prefer, install a floor (or use a concrete slab) and generate the smoke with two or three electric heating elements properly spaced on either end, or perhaps with gas or charcoal heating elements.

The basic A-frame is good for smoking meat and fish on either one, two, or three longitudinal rods, but is not so good for smoking on racks because access and top-of-the-structure space are limited.

COMMERCIAL SMOKERS

You may find commercial smoking units suitable for cold-smoking, rigged with a thermostatically controlled heating element, but they will be rather expensive. Most of the small home smokers, rigged with an electric heating pad, may or may not work for cold-smoking, depending partly on the outside temperature and wind. I've never had a problem with these units

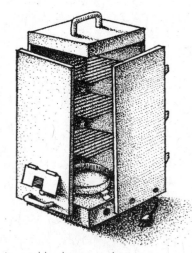

A portable electric smoker

for hot-smoking a few fish, but a whole turkey is another matter. So, proceed carefully—and beware the gray area between 80°F and 140°F.

Although my old Outer's electric smoker has served me well for over thirty years, it is not ideal for cold-smoking, simply because it is difficult to attain enough heat to generate the smoke while at the same time keeping the smoke chamber cool. It's a big problem.

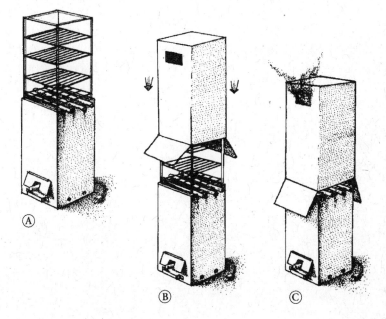

(A) By removing the lid, putting wooden sticks across the top, placing the racks on top of the sticks . . . (B) and fitting the cardboard packing box over the racks . . . (C) you can make a portable smoker taller, and thereby cool the smoke.

One manufacturer, Luhr-Jensen, suggests that you can cool the smoke somewhat by making the portable smoker taller. This trick can be accomplished with top-loading units simply by removing the lid entirely, then putting two wooden sticks across the top of the unit. The rack fits atop the sticks, and the cardboard packing box fits over the racks. The same idea can be used to rig an extension compartment made with wood or tin. Tin would be better for cold-smoking because it conducts heat more readily. Also, you can rig a round extension for the silo smoker-cookers. Thus, you can have a neat rig for hot-smoking as well as for cold-smoking.

Obviously, any top-loading unit will be easier to extend. Further, the extension doesn't necessarily have to go straight up. It can dogleg or make two elbow turns. Just make sure that the smoker draws properly and doesn't get too hot for cold-smoking.

There are some expensive electric smokers on the market with automatic temperature control, and some of these work much better than the smaller units. In my view, however, the culinary sport will want to devise his own smoker. With it he can brag that his smoked meats and fish are better than any other, and the smoker itself becomes part of the secret recipe or at least part of the reason for success. Truly, the improvised cold smoker, however ugly, is a thing of inner beauty and self-fulfillment—and a walk-in rig can be the pride and joy of the patio chef's repertory.

Here are some ideas for the do-it-yourselfers:

STORAGE-SHED SMOKERS

There are lots of inexpensive sheet-metal storage sheds on the market, and custom units are also available. Some of these are

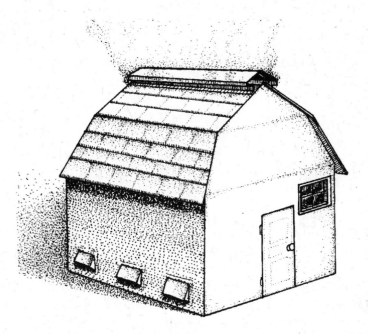

My favorite barn-shaped shed smoker

attractive, but most of them are merely functional. In any case,
these sheds can be used for smoking or be easily modified to do
so. Most people want to put the sheds on concrete slabs, but this
is not a good choice if you plan to generate smoke from a wood
fire. Put the shed either directly on the ground or perhaps on
concrete footers, leaving dirt in the center. If these units are too
tightly constructed to allow the smoke to escape properly, drill
holes in the walls just under the overhang, if there is an overhang. You can also install adjustable vents and possibly even ridge
vents on some units.

Although you can use such a shed for both storage and
smoking, I really don't recommend that practice because the

smoke covers everything and the salt will rust tools and equipment. Instead of using the space for storage, build in a wooden bench for salting and otherwise working with the meat.

While working on this book, I started looking at various storage sheds for sale in my area. The roofs of most of these don't have much slope. My favorite design is shaped like a barn—it has a good high point for the smoke to exit, and the shed I priced even had ridge venting. Perfect. It was attractive, and might well fit into a country setting.

OLD REFRIGERATORS AND FREEZERS

Converting an old refrigerator into a smoker is relatively easy if you've got all the racks. Such a smoker is also convenient to use. Many people put a hot plate in the bottom of the refrigerator and cut a vent hole in the top. This rig may or may not be satisfactory for cold-smoking, depending on how much heat it generates. It's best to check out the unit by leaving a thermometer on one of the shelves, or drilling a hole and installing a more or less permanent thermometer that can be read from the outside.

For cold-smoking in most climates, the source of the heat should be removed from the refrigerator. The heat can be charcoal or electric, or even gas of some sort. The length of the smoke pipe should be great enough to allow the smoke to cool properly, and the slope of the pipe should be great enough for the rig to draw the smoke properly. The rig really works best on a hillside.

If you want to use the heat inside the refrigerator for cold-smoking, remember that the size of the box and the intensity of the heat have a bearing on the temperature. Remember also that these boxes are well insulated, so the outside temperature

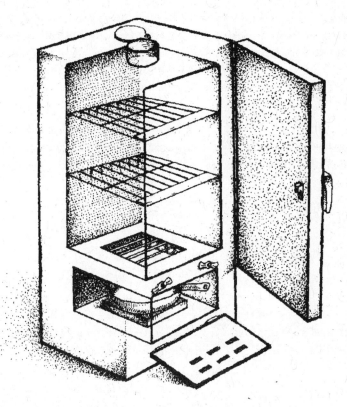

An old refrigerator turned smoker

won't be as important as when using one of the small aluminum smokers that conduct heat rapidly.

In addition to old refrigerators, old top-loading home freezers can be rigged for smoking. These can be fitted with racks and hanging rods.

Be warned that old refrigerators and freezers can be dangerous around small children, and most of them can't be opened from inside. Consequently, the locking handles should be removed or modified so that anyone playing hide-and-seek can get out.

BOX SMOKERS

In Florida, smokers built of wooden boxes with several closely spaced removable trays are popular for smoking mullet and other fish. The trays are made with wooden frames fitted around hardware cloth. These trays slide in and out, and can be rotated from top to bottom during the smoking period.

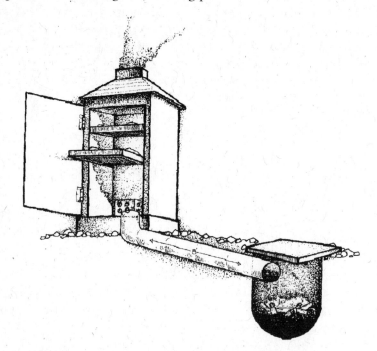

A box smoker with trays.

Some of these boxes sit over the source of heat and smoke, and some have the firebox at some distance from the smoker. The latter is usually necessary for cold-smoking. In either case, the design is great for smoking butterflied fish or such small items as shrimp or oysters. Sausages can also be smoked on the trays.

In addition to the trays, you can also install a pole across the top for hanging hams, or perhaps screw metal hooks into the top.

BARRELS AND DRUMS

Convenient cold-smokers can be made from wooden barrels and metal drums, provided that the fire or heating element is external. Metal drums conduct heat better than wooden barrels, and therefore run cooler, but they are not as quaint as, for example, a used Jack Daniel's whiskey barrel.

If a vertical smoker is desired, cut a hole in the bottom to accommodate a stovepipe or other smoke conduit from the heat source. Then remove the top of the barrel and install racks or bars for holding or hanging fish or meat. Next, fashion a lid for the barrel and drill or cut holes for the smoke to escape. (A smokestack is not necessary, but it will look better if you are setting up a more or less permanent smoker.) Usually, a piece of plywood cut a little larger than the top of the barrel will make a satisfactory lid, and vent holes can be drilled directly into the plywood. Drill only a few holes at first, then add more if needed.

Some people have hot-smokers and -cookers made from old metal drums or barrels cut in half lengthwise, hinged, and fitted with a rack. These can also be used for cold-smoking if you put a vent on the top of one end, and run a stovepipe or other heat conduit to the bottom of the other end. More and more we see large barrel-shaped cookers with small barrel-shaped

A drum or barrel smoker

fireboxes attached to the bottom. Some of these may work for cold-smoking in cool weather. You can sometimes rig an electric hot plate and a sawdust or wood-chip pan in the firebox or a large offset unit, making a cold-smoker from a cooker-smoker. In any experiment of this sort, it's best to make a trial run with a thermometer instead of meat or fish.

So, there you have it. As stated earlier, I suggest that you start cold-smoking with a simple, disposable unit. Then you can expand later, after you determine your individual needs. Much might depend on the design of your house and landscaping. In any case, give some careful thought to the details before building something permanent. Be especially suspicious of the length dimensions given for chimneys and smoke conduits running from the fire to the smoke chamber. When planning smoke conduits, remember that aluminum piping will conduct heat much faster than concrete or plastic pipes. This fact should have a bearing on the required length. I do recommend that you build your own cold-smoker—but proceed cautiously.

BEWARE THE GRAY ZONE

Be warned, again, **that a smoker designed for cold-smoking should not act as an oven and should not function in what I call the gray area.** For cold-smoking, it's best to keep the temperature below 100°F, and preferably at 80°F or less. At higher temperatures, salmonella and other bacteria can multiply rapidly until a temperature of about 140°F is reached. Thus cold-smoking should be accomplished at temperatures below 100°F and preferably at 80°F or lower. Hot-smoking should be accomplished at temperatures higher than 140°F. *Avoid the gray area in between.* Note also that cooking and smoking at 140°F may not be safe in all cases. The U.S. Department of Agriculture is now recommending that your Thanksgiving turkey be cooked to an internal temperature of 180°F—a figure impossible to attain in a 140-degree smoker.

Note also that cold-smoked fish and meat is not cooked. It may or may not be safe to eat, but it is simply not cooked. A good salt cure and proper handling prior to smoking will make the meat safer to eat without cooking, but fully cooking the meat is the safest way to go. Recipes for smoked meats are included in the following chapters. I also recommend using a salt cure for meats that are hot-smoked at low temperatures, and especially in smoker-cookers that may dip below the 140-degree mark in cold or windy weather.

6

WOODS & FUELS FOR SMOKING

Backyard chefs accustomed to heating with a few wood chips placed on charcoal or gas-heated lava rocks can get by with a small bag of chips purchased at the supermarket, where they are available in such flavors as hickory or mesquite or Jack Daniel's whiskey barrels. For cold-smoking in a large smokehouse, however, the practitioner must be concerned with quantity as well as quality, and should have a large supply of suitable wood at hand. It's best and certainly cheaper to cut your own, using whatever good hardwood is readily available to you— pecan, oak, hickory, alder, etc. It is also important to consider the size of the wood pieces, depending on availability and on your cold-smoking rig.

SAWDUST

The sawdust from hardwood makes a dense smoke for a long period of time if used properly. Ideally, it should be smoldered in a pan instead of being put in direct contact with the fire. A cast-iron skillet over an electric burner is ideal. Of course, sawdust can also be piled over hot coals or charcoal. It can be used dry or damp. My favorite is fresh sawdust from green wood. A good supply can be caught in a strategically placed wide open pan while you are sawing your winter firewood.

If you don't cut your firewood, you might visit a sawmill, cabinetmaker, or some tradesman who deals in wood. Many small sawmills simply throw sawdust away, blowing it out onto a pile that will eventually have to be burned or hauled off. But these days more and more sawdust is being used as a by-product, so that you may have to pay for it.

Sawdust from any good hardwood will work, provided that it has not been treated with chemicals.

If you have a walk-in smokehouse, you should consider building a small wood fire on the dirt floor. When this fire burns down to red-hot coals, cover them with sawdust. In about 30 minutes you'll have a dense smoke. You may have to add more sawdust from time to time. If your wood burns out, you may be able to start it again with kindling, but it's probably best to start a new fire outside the smoker, then bring in some hot coals with a scoop. Cover the new hot coals with sawdust and you're in business again.

You can also combine sawdust with chips and chunks.

WOOD CHIPS AND CHUNKS

Various sizes of chips and chunks can be used for smoking, but they are really better suited for use in hot-smoking applications. Buying packaged chips from barbecue supply houses can be quite expensive, but you may be able to find some at a bargain price from large wood-chip operations. In my part of the country, hardwood chips are being hauled out by the trainload.

Dry, soaked, or green chunks can be used in a pan over electric heat, or they can be piled over charcoal or another heat source.

Green or soaked wood chunks can also be used for cold-smoking. These can be piled over hot coals, but green chunks don't work too well in a pan. Again, buying wood chunks from a barbecue supply house gets expensive. It's best to saw off an inch or two of a good log, then chop the wheels into chunks with a hatchet.

You can also cut green limbs into small wheels.

LOGS AND LIMBS

If you have a large walk-in smokehouse with a dirt floor, and have a talent for maintaining a fire, you can use green logs for both the smoke and the heat. This is the old-timey way. It's hard to get green logs just right, but once hot, they will smolder for a long time without making a blaze. A preliminary fire is built in a trench with dry wood and kindling, and sometimes helped along with charcoal. Then the green logs are placed on either side of the trench. As the logs burn, they can be inched closer and closer together. Also, if the trench is properly sloped, the logs will tend to settle down automatically as they burn. With luck, two new logs can be put down and consumed without having to build another fire. Thus, four green logs properly arranged will smolder for days, producing lots of smoke and not too much heat. Almost always, a green, freshly cut log will not burn as hot as a dry log; part of the reason is that the heat is used up in vaporizing the moisture in the wood. Technically, this is called the latent heat of vaporization and explains why boiling water at sea level doesn't get hotter than 212°F.

Smaller logs, cut from limbs or saplings, can be placed directly atop a source of controlled heat, such as an electric burner. A rectangular cast-iron griddle or fish cooker heated over two stove burners can be used to heat green limbs or sticks of stove wood. Again, green wood is my favorite, but dry pieces can also be used if they are merely heated instead of being part of the fuel.

WHERE THERE'S SMOKE . . .

In cold-smoking, you'll need enough heat to produce smoke—but not enough to overly heat the smokehouse or smoke chamber. Obviously, much will depend on the size of the smoke chamber, the location and intensity of the fire, and the outside temperature.

- *Electric Heat* The familiar hot plate, usually with one or two coil-type rheostatically controlled heating elements, is by far the easiest way to go for most small-scale cold-smoking operations. They are quite efficient because a wood-chip or sawdust pan can sit directly on the coils; in short, they can easily heat the wood to the smoke point without heating up the smoke chamber too much. Some of these can be purchased from sausage and smoking equipment suppliers, but these are often expensive. You should be able to rig your own, using an old hot plate or a new one from a discount store. I recommend a cast-iron skillet to hold the chips or sawdust. Some of the small commercial "smokers" have a tin pan with a wooden handle.

  A single or double heating element will generate enough smoke for most chambers, but a walk-in unit may require two or three hot plates. An old electric kitchen stove or stovetop will also work if it is properly installed (these usually require 220–240 volts, whereas most hot plates operate on 110–120 volts).

- *Gas Heat* Any small gas burner can be used much like an electric hot plate, and portable units with gas cylinders will work where electricity isn't available. The larger refillable tanks are better (and cheaper in the long run) than small gas cylinders. Also, natural gas can be used if you have it. Of course, the burner should have a valve to adjust the flame. It's best to use a cast-iron skillet to hold the wood chips or sawdust for these units.

- *Other Burners* Any sort of camp stove, such as Sterno, can be used for generating smoke in a skillet or wood pan. You'll need a long-burning stove for cold-smoking, unless you can be at hand to add new fuel as needed.

- *Coals* You can heat sawdust and wood chips with hot coals, produced by burning wood, charcoal, or hard coal. Usually, the sawdust is piled onto and around the coals. The danger is that the sawdust or chips will become fuel for the fire, quickly burning up and raising the temperature in the smoke chamber.

- *Wood* It is possible to combine wood for fuel and wood for smoke in large smokehouses with a dirt floor or, sometimes, in units that have the fire a good ways from the smoke chamber. For this purpose, I prefer to start the fire with dry wood, then add some freshly cut green wood for smoke. As the green wood burns into coals, more green wood can be added, thereby keeping the fire going and the smoke coming, if all goes according to plan. Parallel logs can be used in walk-in smokehouses, as pointed out earlier in this chapter. Be warned that this method requires lots of attention, but also remember that letting the fire go out from time to time isn't disastrous, except possibly early in the process when blowflies or other insects are a problem.

Personally, I find this method the most satisfying way to cold-smoke meat and fish. I believe that the quality of the smoke is better than smoke generated in a pan, but I have no explanation for the difference, be it real or imaginary.

- *Combinations* You can combine the methods above, using, for example, green wood logs during the day and electric hot plates during the night.

THE BEST WOODS FOR SMOKING

Although some people champion one wood or another for smoking meats, fish, and game—often saying to use one wood for red meat and another for fish—I don't put too much stock in their claims, and I wouldn't hesitate to use any good hardwood that is readily available. For either hot-smoking or cold-smoking, I almost always prefer freshly cut wood to seasoned wood, but again, there are opinions on this matter. If you prefer green wood, you'll have to cut your own or make a deal with a local wood-cutter. The little bags of wood sold by barbecue supply houses are dry. I think that cutting your own wood is part of the fun. A chain saw is handy, but a good bow saw will also cut lots of good wood in short order.

In most parts of the country, suitable woods can be gathered easily from woods and fencerows. Here are a few favorites:

- *Hickory* Most of the country people in my neck of the woods use hickory for cold-smoking. That's what my father and grandfather used. Hickory does indeed make very good smoke, but I think its reputation was built on the long-burning qualities of the green wood. What the old-timers wanted was a wood that would smolder all night.

    A friend of mine asked me how to identify a hickory tree. Look for remnants of nuts that squirrels have gnawed scattered around base of the tree. I have been told that the nuts can also be used for smoking, but, it seems to me, these would be better for hot-smoking where only a few would be used during the cooking process.

- *Alder* This wood is quite popular for smoking fish in the Pacific Northwest and some other areas, and alder chips and chunks are packaged and sold all over the nation. Some practitioners who prefer alder go so far as to say that it's the only smoke to use for fish and poultry. I'll allow that some of the best salmon I've eaten was smoked with alder, but it was smoked with freshly cut green alder. If you have plenty of alder free for the cutting, fine. If not, use some other good hardwood and lie about it.

- *Guava and Other Fruit Woods* The guava tree grows in South Florida and other tropical or subtropical regions. It is especially popular as a smoking agent in Hawaii. Citrus, plum, cherry, pear, and other fruit woods can also be used.

- *Pecans* These are my personal favorite for both cold- and hot-smoking, partly because I grew up in pecan country. The nuts fill out in the fall of the year and the tree limbs get heavy. A high wind, and sometimes ice, can cause large tree limbs to break off, especially if they are heavy with green nuts. These limbs are ideal for smoking, and most grove owners are glad to let you to haul them off. Of course, wild pecan trees also make excellent wood for smoking.

  If you live near a large pecan sheller, consider trying pecan shells instead of sawdust in a smoke pan over an electric burner. The shells are sometimes marketed for this purpose in expensive little bags, but you may be able to get shells free from a sheller.

- *Sassafras* This tree often grows quite abundantly on fencerows in my part of the country. The small trees, from 4 to 6 inches in diameter, are easily felled and easily handled, and make good logs for burning in a walk-in smokehouse. Sassafras wood is easy to cut when green, but when it seasons it becomes very hard. It was once used for railroad cross ties and in making chicken coops.

  When you cut some sassafras saplings for smoking purposes (or when you pull some roots for herb tea), also pick a few green leaves. Dry these, crush them in a mortar, sift, and use the powder (very sparingly) to thicken soups and gumbos. Powdered sassafras leaves are a Native American seasoning and thickening agent, marketed at high prices under the name of *filé powder* in Cajun markets and in the spice sections of most large supermarkets.

• *Beech* Wood of the beech tree is used for smoking in some areas, and at one time it was popular in France for cold-smoking herrings.

• *Mesquite* This tree has always been popular in parts of the Southwest, where it grows in very dry areas. It is highly regarded as fuel when other trees are scarce. Owing to its demand on the patio, mesquite is now bagged and shipped to Florida, Maine, Alaska, and points in between. The smoke is good, but the wood's early reputation was earned as a fuel, not as a smoking agent. It makes a hot fire with long-burning coals, which is great for camping and chuck-wagon cookery. Unless you have a ready supply of mesquite free for the cutting, I wouldn't recommend buying it in large quantities for cold-smoking. But I wouldn't make that statement in Texas, west of the Pecos.

• *Mangrove* Wood of the mangrove tree is very good for smoking. Its high reputation for smoking mullet in Florida became an environmental concern. It is now illegal, at least in some areas, to cut or otherwise harm mangroves.

- *Roots* Palmetto roots are highly touted in parts of Florida. Recently my wife ate some ham from a feral hog bagged near Sopchoppy, Florida, and smoked locally with a combination of palmetto and hickory. She says it was the best she has ever tasted. Out West, manzanita roots are highly touted.

- *Maple* This good wood is popular in some quarters. According to A. J. McClane, maple or apple is "the only wood" to use for smoking clams or oysters. He doesn't say why.

- *Oak* Although it is used to flavor some expensive Scottish smoked salmon, which I understand is eaten in Buckingham Palace, as well as for smoking hams in England, oak is really not very popular in America as a smoking agent. Ideal for a campfire, oak produces good hot coals but not much smoke as compared to other hardwoods. In any case, oak can be gathered freely in most parts of the country, and I recommend it highly when it is green, when it is easy to cut and split. Green oak doesn't burn quickly, which makes it a good choice for walk-in smokehouses, where parallel oak logs can be helped along, if necessary, with charcoal or a little dry wood.

- *Pimento, Myrtle, and Juniper* The wood of the allspice tree, called pimento, is used for smoking jerk in Jamaica. The tree is a kind of myrtle—and evergreen—which is generally considered to be a no-no in smoke cookery. But remember that myrtle is also used to smoke meat on the island of Sardinia. I also understand that juniper has been mixed with other woods to smoke salmon on a large scale.

- *Driftwood* I have read that driftwood makes a good smoking agent, but I don't understand how it could be consistent unless you can distinguish oak driftwood from that of hickory or eucalyptus. Maybe it doesn't matter.

- *Other Wood and Combinations* Hardwoods such as walnut, maple, chestnut, sweet bay, ash, or birch can be used to advantage for making smoke. Of course, some culinary sports and advertising specialists will hold out for local favorites, and some discriminating practitioners will say that a combination of certain woods is the best way to go. A fellow from South Carolina, for example, has said, "I prefer smoking with one-third hickory, one-third oak, and one-third fruit tree (plum)." Who could argue with that?

   Dr. Eph Wilkinson notwithstanding, I'll have to add here that my father would have let out a belly laugh if somebody had suggested that he mix sassafras with his beloved hickory.

- *Corncobs* Back when local gristmills and home (or farm) corn shellers were everyday equipment, cobs were often used for fuel as well as for smoking meats—and for other activities conducted in small outdoor structures. I'm sure that purists will have opinions on whether the white cob or the red is better for smoking. In any case, cobs tend to blaze up and should be watched carefully.

- *Peat and Organic Smokes* Peat is used as a smoking agent on some of the British islands. Of course, one sort of peat is not as good as another for smoking purposes. Also, remember that in some barren lands such as Iceland sheep chips are used for smoking.

7

COLD-SMOKED FISH

Be warned once again that salt, not smoke, is the curative agent in cold-smoked fish. In addition to inhibiting the growth of harmful bacteria, the salt draws out moisture—and moisture is necessary for bacteria to thrive. How much salt and for how long depends on the size of the fish and other factors, and how long the fish is to be preserved. As a rule, the more salt, the drier the fish and the longer its shelf life. In modern times, the trend has been toward less and less salt combined with refrigerated storage, and this chapter tends to reflect this change. (If you want completely cured smoked fish, begin with a hard salt cure as described in Chapter 2 and cold-smoke the fish in the smokehouse for a week or longer.) The trouble with a light cure–light smoke approach is that the practitioner has no hard and fast rules to follow. In any case, I can only hope that I can shed enough light on the subject to keep the novice from proceeding in the dark.

DRESSING AND HANGING THE FISH

It's best to start with very fresh fish. If you catch your own, gut and ice them as soon as possible. Better, gut them and put them into a slush of ice and brine; this will start the salting process right away. Before gutting and cleaning the fish, however, consider how they will be handled for salting, drying, and smoking.

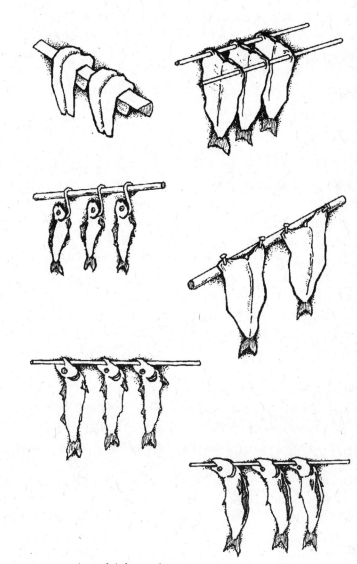

Common ways to hang fish for smoking

Fish to be smoked can be suspended vertically in several ways. They can also be placed horizontally on a rack. How they are to be hung depends partly on practical considerations, such as your facilities as well as the size of the fish. Here are some basic techniques to consider:

Gibbing

This is a method of dressing small fish without slitting the belly open. Instead, a small cut is made across the bottom of the fish, close to and behind the gills. The gills and innards are pulled out and the body cavity is washed thoroughly with cold water. (A hose with an adjustable, pistol-grip nozzle helps do a good job.) After gibbing, the fish can be hung on a rod, running it through the mouth and out the gill, or vice versa. They can also be hung by pushing the rod through the eye sockets. In either method, it's best to leave a little space between the fish.

Gutting

Small whole fish are usually dressed by cutting a slit in the belly and removing the innards. I prefer this method to gibbing. Gutting may or may not take a little longer, depending on how thoroughly you want to wash out the cavity. The fish can be hung by the methods described under gibbing.

Butterflying

This method of dressing a fish leaves the two halves hinged at the belly. First, the fish is beheaded. During beheading, the collarbone is left in if the fish are to be hung; it can be removed if the fish are to be placed on a rack. After beheading, the fish is cut from the head to the tail from the top. Cut as close as possible to the backbone and through the rib cage. (With large fish, you may prefer to work the knife around the rib cage, leaving the ribs attached to the backbone.) It is important that the knife not cut all the way through the belly flesh. Then the fish is turned and cut in a similar manner on the other side. Next, the backbone and tail can be lifted out, along with the innards. Finally the fish is opened like a butterfly. It can be hung or placed on racks.

Filleting

Large fish should be filleted for both curing and smoking. In most cases, a slab of fish is cut off either side of the backbone. Of course, the process should waste as little meat as possible. In my opinion, it's best to start the cut from the tail, then cut through the ribs as you approach the head; this will leave the rib bones intact. For completely boneless fillets, you can cut out the rib section or remove the bones one at the time. There are other methods of filleting, and some people may prefer to work the knife in at the head and toward the tail. In either case, leaving the skin on the fillet will help hold it together during curing, smoking, and handling.

For smoking, fillets are usually placed skin side down on suitable racks. They can also be draped across a rod, skin down.

Tailing

Small whole fish, such as smelts, can be gibbed or gutted and tied at the tail in pairs. These can be hung over rods for smoking. One pair should not be allowed to touch another.

THE SALT CURE

As a rule, it's best to dry-cure fish with a low fat content, such as cod, and brine-cure fish with a high fat content, such as mackerel. But the methods are usually interchangeable, especially if you are after a rather light cure to be followed by refrigeration or freezing. Unfortunately, there are infinite variations on the basic methods of salt-curing fish in preparation for smoking; instead of rhyme and reason, we have purely personal and regional variations, which are often staunchly defended. I'll try to keep it simple.

The Brine Cure

Although many ingredients are sometimes used in the cure to flavor or preserve the fish, most of these are not necessary and may be counterproductive. I do, however, insist on either brown sugar in a dry cure or molasses in a brine cure. The sugar or molasses helps the color as well as the flavor of smoked fish. The measures can be increased or decreased proportionally. Note that these proportions yield a relatively weak brine, as fresh water will dissolve well over 2 pounds of salt per gallon. Note also that water is a major ingredient in the brine cure, and should have a good, clean, fresh taste. Distilled water isn't usually necessary, but some water tastes of natural elements, such as sulfur, and sometimes of added chemicals, such as chlorine. If in doubt, use distilled water or perhaps bottled mineral water. Water from a good spring is hard to beat.

1 pound salt
1 gallon good water
1 cup dark molasses

Dress the fish and place them in a nonmetallic container, then cover with the brine, following the proportions above. Place a plank or plate over the fish and weight it with a bowl, stone, or some such nonmetallic object. The idea is to completely submerge the fish. Leave the fish in the brine for 12 hours or longer. (For a hard cure, more time is recommended, as was discussed in Chapter 2.) Remove the fish from the brine, but do not wash. Dry the fish and smoke them, as described later in the chapter.

The Dry Cure

The dry cure should be started by first soaking the dressed fish in a light brine made with 1 cup salt dissolved in 1 gallon water for a period of about 1 hour. (You can actually start this initial brining as you dress the fish, timing the process after dropping in the last one; if I have a choice, I dress the larger fish first simply because they should have more time in the brine.) After soaking the fish, drain them but do not fully dry them. Have a dry cure ready, using the following 8-to-1 formula:

    8 pounds salt
    1 pound dark brown sugar

Mix the salt and sugar. Place part of this cure (about ½ inch) into the bottom of a curing box, then put the rest into a handy container. Place each fish in the cure, coating both sides, and pick it up with as much cure as will adhere to it. Place the fish on top of the layer of cure in the box. Layer the fish if necessary, sprinkling extra cure between each layer. Leave the fish in the curing box for about 12 hours, or longer for whole fish that weigh more than 1 pound.

Wash the salt cure off of the fish, and then dry the fish as described next.

Forming a Pellicle

After the fish are treated with either a brine or a salt cure, they should be washed and dried before putting them into the smoker. The drying helps in a number of ways, but basically it forms a pellicle, or glazed film, on the surface. This film helps give the fish a better and a more uniform color, and it may also aid in preservation.

After washing the salt from each fish, pat each one dry with paper towels and place it on a rack (or hang it) in a cool, breezy place. If you don't have a natural breeze, a fan will help. Usually the pellicle will form within 3 hours or less, depending on the size of the fish. After the pellicle forms, proceed with the cold-smoking operation.

Cold-Smoking

The first requirement for cold-smoking is to keep the temperature in the smokehouse or smoke chamber cool. Usually, this requires having the fish a good distance from the fire or source of heat, as discussed in Chapters 5 and 6. It's best to keep the temperature below 90°F, and certainly below 100°F. I consider 70°F to 80°F ideal.

If the salt cure was sufficient to be on the safe side, you can cold-smoke the fish for a few hours, a few days, or a few weeks. As a general rule, the longer the fish is smoked, the more smoky the flavor; but at some point the law of diminishing returns sets in and the smoking, if done in a dry atmosphere, becomes more of a drying period than a flavor-enhancing period.

For the modern practitioner, I recommend that the fish be smoked until the surface of the flesh is a good mahogany color and the texture is firm but still pliable to the touch. Drying out the fish is unnecessary if you are going to cook them right away. They can be refrigerated for a week or so, or wrapped and frozen for several months. If you do dry the fish, remember that they should be freshened overnight in cool water before they are cooked.

It is best (but not necessary) to start cold-smoking fish with a light or moderate smoke, then increase the density. It's also best to keep the smoke coming continuously, but all is not lost if your fire goes out for an hour or so or the fuel gets low in a chip pan over electric heat. Merely resume smoking as soon as you can.

How long should the fish be smoked? A day or two with whole fish of about 1 pound will be about right, but this statement has to be qualified with a bunch of "ifs," "ands," and "buts." The real concern in my conscience is that the fish not be salt-cured properly and then be exposed to a temperature high enough to promote rapid growth of harmful bacteria (higher than 80°F)—but not high enough to cook or retard the bacteria (140°F). Again, beware the gray zone.

How long? If the fish is to be cooked right away, they should be smoked at least long enough to give them flavor—at least several hours, preferably 12 hours or longer for whole fish of 1 pound or more. If the fish are to be preserved for 2 weeks, they should be properly salt-cured and smoked for at least 24 hours; 4 weeks, 48 hours. If the fish are to be preserved indefinitely, they should receive a *heavy* salt cure and be smoked for several weeks.

In short, if you first salt-cure your fish, cold-smoking is easy if you have a suitable smokehouse. It may be hard to duplicate a particularly good batch exactly (unless you measure all the variables scientifically and keep meticulous notes), but this isn't an overriding concern unless you are smoking fish commercially.

My best advice is to salt-cure your fish, then cold-smoke more or less at your convenience, give or take an hour or two either way, until the fish have a good color and texture. Then either cook the fish right away, refrigerate them for several days, or freeze them for several weeks. Simple enough? If not, you may want to follow the more specific recipes below, some of which reflect the philosophy of smoking mainly to add flavor instead of preservation.

MODERN COLD-SMOKING METHODS

Although thoroughly cured fish are normally smoked for a long period of time, partly to dry the flesh, you can cut back on the time if the product is to be cooked right away. The shelf life of partly smoked fish can be extended by refrigeration or by freezing. Here are some suggestions for cutting back on the salt or on the smoke, or both.

2-Day Smoked Salmon

This salmon method calls for a relatively short cure and short smoking time. After smoking, the fish should be cooked by some method of your choice, or used in the recipes set forth later in this chapter.

1 very fresh salmon, about 10 pounds
1 gallon good water
2 cups Morton Tender Quick Mix
1 cup light brown sugar

Fillet the salmon and rinse it. Mix the Tender Quick and brown sugar into the water. Put the salmon fillets into a nonmetallic container and cover them with the curing mixture. Place a plate on top so that the salmon fillets are completely submerged, then place the container in the refrigerator for about 24 hours. Remove the salmon fillets and air-dry them. Then arrange the fillets on your smoker racks skin side down, or drape them over a cross beam, skin side down. (Do not attempt to hang the fillets by the end. If you use the rack in a small smoker, you may have to cut the fillets in half.) Cold-smoke for 24 hours or so.

Then cook the salmon as described below, or use a recipe such as kedgeree. I like to freeze one fillet for later use, and steam the other fillet for 15 to 20 minutes, or until it flakes easily when tested with a fork. To steam the fillet, cut it in half and place it on a plate, skin side down. Put the plate in a steamer for 15 minutes, or until it flakes easily when tested with a fork. Serve the steamed salmon skin side down with rice and vegetables of your choice.

To bake the smoked salmon fillets, simply put them skin side down on a suitable flat baking sheet and bake in a preheated 300°F oven for 25 minutes, or until they flake easily when tested with a fork. Baste once or twice with butter or bacon drippings while baking. Do not overcook.

Cold-Smoked Shrimp

Here's a method for cold-smoking precooked shrimp. It will add flavor, but the shrimp should be refrigerated after smoking and should be eaten within a week. For best results, you'll need a large pot that won't crowd the shrimp during boiling. Add some salt—about 1 cup per gallon. Remove the heads of the shrimp and devein them, but leave the tails on. Bring the water to a boil, add the shrimp, bring back to a boil, and cook for 3 minutes for medium shrimp. (Large shrimp require another minute; smaller shrimp, less time. The shrimp are ready when they turn pink.) Drain the shrimp and arrange them on racks. Cold-smoke for 30 to 60 minutes, depending on size. Eat the smoked shrimp cold, or put them into shrimp salad or other dishes. I like a few chopped shrimp scrambled with chicken eggs and a spring onion or two chopped with half the green tops.

Note: Many people add all manner of Cajun spices and court bouillon vegetables to a shrimp boil. I think that plain salt or sea salt is better, but suit yourself.

Smoked Oysters

At the time this was written the Food and Drug Administration was considering a ban on the sale of oysters on the half-shell. It's hard for me to imagine doing without oysters in the fall. I still eat them, and will continue to do so as long as I can buy them from a reputable dealer. I don't know what led it to consider the ban, but my hope is the FDA would make certain of its facts before prohibiting such a delicacy. (Sometimes government agencies do crazy things, such as trying to ban lead fishing sinkers under ½ inch in diameter on the grounds that they harm birds. A sinker of this size might do harm to a seagull when slung out by surf casters, but that's the only way.) My concern about banning oysters on the half-shell is that it will surely kill the method of shipping fresh oysters in this country. When taken alive, they can be put into burlap bags and shipped in refrigerated trucks. When the consumer buys them, the oysters are still

alive and still contain the salt water from the bay. Thus, they can
be the freshest of all seafoods, along with clams and mussels. If
they are shucked and put into containers before shipping, they
will be dead and rather milky in color, and will lack the flavor
of the sea.

In any case, freshly shucked oysters can be smoked successfully, but I really can't recommend the preshucked "bucket oysters" for sale in supermarkets. Most of these have been washed,
or otherwise lack the flavor of their natural juices. An oyster
should never be washed, George Leonard Herter notwithstanding, even if it is to be cooked in a stew.

For smoking oysters, you'll need fine-meshed racks so that
they won't fall through. Grease the racks and build up a head of
smoke. Shuck the oysters and place them on the racks without
touching them. (Any oyster that is not alive should be thrown
out. Also throw out any oyster that has opened, and any oyster
that seems a little dry as compared to the others.) Smoke the
oysters for 1 hour, or until they have taken on a nice golden
color. Place the smoked oysters in a container and coat them
with olive oil. Refrigerate and eat within a few days.

If you are squeamish about eating raw oysters, try steaming
them on a grill over a hot fire until they pop open. Then shuck
them and smoke for 30 or 40 minutes.

A. D.'s 1-Day Wonder

Early in the morning, catch some good fish of about 2 pounds
each. Fillet the fish. Soak the fillets for a couple of hours in a
brine made of 2 cups salt per gallon of water. Dry the fish in
a breezy place for 2 or 3 hours, or until a nice pellicle forms.
Cold-smoke the fish for about 6 hours.

Fry some smoked bacon in a large skillet. Remove the bacon and save it to eat with the fish. Reduce the heat to the lowest setting and sauté the smoked fillets in the bacon grease, turning them from time to time. If your skillet is large enough, also sauté some sliced portobello mushrooms. If necessary, use a separate skillet for the mushrooms. For a complete meal, serve with white rice and steamed vegetables.

Of course, this technique also provides a recipe for cooking the fish after smoking. Here are some other recipes to try with smoked fish:

RECIPES

Although I have eaten uncooked cold-smoked fish, I recommend that they be cooked by a suitable method. Here are some suggestions.

Scrambled Smoked Fish

This dish is one of my favorite ways of using a small amount of smoked fish. It's also one of my very favorite breakfast dishes. Be sure to try it with either cold- or hot-smoked fish. I use what I call spring onions, which most cookbook copy editors want to change to green onions. They can be purchased all year in supermarkets. About the size of a pencil, they are normally sold in bunches. For this recipe, I use about half the green tops along with the small bulbs. First, I cut off the root ends and the tips of the green parts. Then I peel off a layer and chop what's left. I also use wild onions with green tops, but be warned that some of these are quite strong.

1 cup flaked smoked fish
6 large chicken eggs

1 tablespoon milk
2 tablespoons butter or margarine
2 or 3 spring onions with tops, finely chopped
salt and freshly ground black pepper to taste

Break the fish into small pieces, or chop it with a chef's knife. Beat the eggs with a little milk, then mix in the fish flakes and chopped green onions. Heat the butter in a skillet, add the egg mixture, and scramble until the eggs are set. Sprinkle with salt and pepper.

This makes a wonderful breakfast, especially when served with slices of luscious red homegrown tomatoes and fresh toast.

Russian Salmon and New Potatoes

Here's a recipe that I adapted from Kira Petrovskaya's Russian Cookbook. *Although billed as a salad, it makes a nice lunch for two people on a hot day. The olives can be either green or black, or a mixture. Black ones, however, make a more attractive salad.*

½ pound smoked salmon
4 or 5 boiled small new potatoes, diced
½ cup sliced pitted olives
1 tablespoon minced onion
1 tablespoon minced green onion, with some tops
1 tablespoon drained capers
1 tablespoon wine vinegar
2 tablespoons olive oil
1 teaspoon prepared mustard
salt and pepper

Mix the onion, green onion, capers, olives, and potatoes in a bowl. Cut the salmon into fingers and mix very carefully into the potato mixture. Chill.

Prepare a dressing by combining the vinegar, oil, and mustard; season with a little salt and pepper to taste. When you are ready to serve, pour the dressing over the salmon, but do not stir in. Eat immediately with crackers or toast.

Smoked Fish Patties

Here's a dish that I like to cook on a cast-iron griddle. The fish patties can be eaten as part of a meal or used in a sandwich. For the latter, all you'll need in addition to the patties is some mayonnaise generously spread onto each slice of bread.

2 cups flaked smoked fish
1 cup fine cracker crumbs
3 green onions, minced with about half the tops
½ red bell pepper, seeded and minced
1 clove garlic, minced
2 medium chicken eggs, whisked
salt and pepper to taste
flour
butter

Mix the fish, cracker crumbs, green onions, red pepper, garlic, eggs, salt, and pepper in a bowl. Shape the mixture into patties, then dredge each patty lightly in flour. Heat a little butter on your griddle or in a skillet. Over medium heat, sauté the patties for a few minutes or until nicely browned on the bottom, then turn carefully with a spatula and brown the other sides. If the patties tend to tear apart, use two spatulas, one on top and one on bottom, for turning.

This makes 4 large patties. I need 2 for a full meal, but can get by with 1 for lunch or in a sandwich. I like to sprinkle the patties with oriental fish sauce (*nam pla* or *nuac mam*), which can be put on the table as a condiment. Oyster sauce is also good.

Avocado Stuffed with Smoked Fish

The best avocados I have ever eaten were grown on a large tree near Tampa Bay. This area is also noted for its smoked mullet. The two ingredients fit nicely together in a recipe from Ghana, which I have adapted here. Any good smoked fish will do, but those that have been cold-smoked should first be steamed or poached for about 15 minutes, or until they flake easily when tested with a fork.

```
1 heaping cup flaked cooked smoked fish
2 large avocados, quite ripe
4 chicken eggs, hard-boiled
¼ cup olive oil
¼ cup peanut oil
¼ cup milk
2 limes
½ teaspoon salt
¼ teaspoon sugar
```

Peel the whites off the egg yolks. In a bowl, mash the yolks with the milk until you have a paste. Add salt, sugar, and the juice of 1 lime. Stir in the peanut oil a little at a time, then stir in the olive oil. Chop the egg whites and add them to the bowl. Stir in the fish. (This stuffing can be refrigerated for later use, or served at room temperature.)

When you are ready to eat, cut the avocados in half, remove the pits, and fill the cavities with the stuffing. Squeeze on a little lime juice and enjoy.

In Ghana, the dish is served with strips of red bell pepper or pimiento as a garnish. This is a nice touch. I also like it sprinkled with paprika and served on lettuce, along with wedges of lime so that each person can squeeze some juice onto his serving.

Note: This stuffing makes a good spread for crackers, especially when made with garlic oil (olive oil in which garlic cloves have steeped for several months; I preserve garlic in this way, then use the oil for cooking purposes).

Irish Herring

According to my old edition of *Larousse Gastronomique,* the Irish once had a simple but unusual method of preparing smoked herring. The fish were beheaded and split in half lengthwise, spread flat in a suitable container, covered with whiskey, and flamed. When all the whiskey had burned away, the herring were ready to eat. I recently ran across the same technique in a book on British country cooking, in which the author used the term *Rob Roy* for cold-smoked fish prepared by this method.

One day I intend to try the method with real Irish whiskey and smoked herring, but to date I have not had the opportunity. I did, however, prepare some at a smoked mullet fest in the north of Florida, where the tradition is to butterfly the mullet before smoking it. I selected some smaller smoked mullet and poured some sour-mash bourbon over them and burned it off. The fish were delicious. The idea got some attention in the area, and the North Florida good ol' boys may have picked it up.

Smoked Fish with Cream

Here's a dish that I like to cook in a large cast-iron skillet. A large electric skillet will also work.

    2 pounds smoked fish
    ½ cup half-and-half or light cream
    2 tablespoons butter
    salt and freshly ground black pepper
    chopped fresh parsley or green onion tops

Melt the butter in a skillet and sauté the fish for 1 minute on each side. Sprinkle on a little salt and black pepper, then pour in the cream. Simmer (do not boil) for 5 minutes. Turn the fish and simmer for another 5 minutes. Place the fish on a serving platter and pour the pan liquid over it. Garnish with parsley or green onion tops. Serve with boiled new potatoes, steamed vegetables, and hot bread.

Kedgeree

This dish from Britain takes its name from the Indian khichri, *but one English writer suggests that the only thing Indian about it these days is the curry powder. In America, even that ingredient is sometimes omitted, thank god.*

    ½ pound smoked fish, cooked and flaked
    2 large onions, chopped
    ½ cup butter or margarine
    1 hard-boiled chicken egg, sliced
    4 tablespoons light cream or half-and-half

½ lemon
1 cup long-grain rice
salt (if needed) and black pepper to taste
1 teaspoon curry powder (optional)

Bring 2 cups of water to a hard boil, add the rice, return to a boil, reduce the heat, cover tightly, and simmer for 20 minutes without peeking.

If the smoked fish has not been cooked, steam or poach it until it flakes easily, about 10 minutes. Heat half the butter in a large skillet and sauté the onions for 5 minutes. Add the fish, stirring about, and mix in the curry powder, pepper, and salt, if the latter is needed (depending on the salt content of the fish). Add the sliced egg.

Melt the rest of the butter and mix it into the rice, which should be fully cooked by now. Stir the rice into the fish mixture. Squeeze the juice of half a lemon over the kedgeree, then spoon the cream over all. Serve hot. Feeds 3 or 4.

Smoked Fish Chowder

I've never cooked a chowder that wasn't good (at least to me), but I'll have to admit that I am partial to one made with smoked fish. Here's an old recipe from Maine. Any good smoked fish can be used, hard- or light-cured, but fish with large scales should be scaled before being boiled.

1 whole smoked fish, about 1 to 1½ pounds
¼ pound salt pork, diced
3 or 4 medium to large potatoes, diced
1 cup diced onion

2 or 3 cups whole milk
black pepper to taste

Put the fish in a pot, cover it with water, bring to a boil, reduce the heat, and simmer until the fish is tender. (The time will depend on how hard it was cured and how long it was smoked.) While simmering the fish, sauté the salt pork in a skillet until the pieces begin to brown. (Some cooks may want to put the salt pork in with the fish instead of browning it; suit yourself.) Drain the salt pork and sauté the onion in the skillet for 5 minutes.

When the fish is tender, remove it from the pot, and flake the meat, discarding the bones. Add the potatoes, salt pork, and pepper to the pot. Cover and simmer for 10 minutes. Then add the onion and simmer for another 10 minutes or so, or until the potatoes are done. Stir in the fish flakes. Slowly add from 2 to 3 cups of milk, stirring and tasting as you go. Ladle the chowder into bowls and serve with plenty of hot bread. A loaf of chewy-crusted French or Italian bread suits me better, but hot biscuits or even bannock will do. I like to grind some more black pepper onto my serving. Feeds 4.

8

COLD-SMOKED MEATS

Most meats that have been properly cured can also be smoked. It is so simple, once you have the facilities and the skill to maintain the smoke at about 80°F. Since the meat will be fully cooked after it is smoked, you can cold-smoke it for as long as you choose, within reason. Ideally, it should be smoked just long enough to get a good flavor but not long enough to dry out the meat. If you are smoking the meat for long or unrefrigerated storage, however, it will be necessary to dry it out somewhat. If you plan to cook the meat soon after smoking, color is a good indication of readiness. Most nicely smoked meats and fish take on a mahogany brown color. But color alone is not a foolproof measure of the quality of smoked meat, since many commercial operators use dye in or on the meat.

The first step in cold-smoking meat is to cure it with salt by either the dry-salt or the brine method. These subjects were treated at length in Part One. At the risk of boring the reader, I repeat some of the cures, or variations, in the recipes below. My fear is that some no-salt freak will omit this important step, or cut back on the cure, thereby making the smoked meat danger-ous. Even with plenty of salt, most cold-smoked meats should be thoroughly cooked before you eat them, although raw smoked meats, usually thinly sliced, are considered to be delicacies in some quarters. I have eaten uncooked meats from time to time, and I have no objection to the practice if I am certain that the meat was handled properly before and after curing and smoking. I have even eaten Middle Eastern *kibbeh,* made with raw ground meat, but I can't recommend the practice to others—and certainly not with supermarket fare these days.

COLD-SMOKED HAMS

If you have salt- or sugar-cured a ham successfully, the smoking part is easy. It can be aged and then smoked, or it can be smoked first and then aged. I prefer the latter method. (The smoking period can be considered to be part of the total aging time.) After the salt equalization period, wash the ham with warm water, then hang it in the smokehouse by the shank to dry for several hours or, better, overnight. Since the ham is (or should be) completely cured, the temperature is not as critical as for some other smoking operations.

The cured ham can be cold-smoked for 2 days or several weeks, depending on your schedule and on how hungry you are. I recommend a minimum of 2 weeks for cold-smoking. If the smokehouse is screened, it is not necessary to keep the smoke going for the entire 2 weeks. If the fire goes out for a few hours, no harm will be done.

It's best to hang the hams by the shank end, using strong cotton string. My father used withes of green hickory for tying the hams; my wife's father used strips of yucca leaves, called bear grass in her neck of the woods.

The smoked and aged ham will be quite hard and should be cooked pretty much like the salt-cured hams described in Part One (or by one of the recipes below). If you don't want to cook a whole ham, have it sliced into steaks, saving the smoked hocks for such dishes as hoppin' John. The steaks should be soaked overnight in water before cooking, unless they are going into a stew.

Smoked hams to be aged can be put into muslin bags and hung with a string tied around the bag at the shank end of the ham. For security, it's best to fold over the top of the bag before wrapping the string around it. Some people paint the bag with a wash made with various materials, including lime, clay, and flour. Others dust the surface of the ham with a mixture of finely ground red and black pepper. Unless insects are a problem, I think the open-air method is best, provided the ham can be hung freely and in a cool place. If you don't have a bag, a clean, well-used cotton sheet will do. Tear out a square, center the butt of the ham on it, bring the corners together over the shank, and tie off tightly.

Cooking a Cold-Smoked Ham

A home-smoked, cured ham is vastly different from—and better than—a supermarket cured ham. At least to my taste. But it *must* be freshened and cooked differently. Usually, the result will be much firmer than a supermarket ham, some of which even have a spongy texture. In most cases, a cured ham should be sliced very thin before it is served.

In addition to the recipes below, those set forth for salt-cured hams in Part One can also be used for smoked hams.

Dr. Eph's Lament

Before setting forth his recipe in A Man's Taste, *published by The Junior League of Memphis, Dr. Eph Wilkinson said, "The quality of the ham makes all the difference in any country ham recipe. I grew up on a farm in East Tennessee where we killed hogs every fall. The hams were sugar-cured and smoked with hickory and sassafras. I sure wish I could get just one of these hams!" I too would like to have one of those hams, but I tested Dr. Eph's recipe with an ordinary salt-cured and smoked ham of my own devising, and I found it to be a very good one. I'm not sure what the Coca-Cola does, but it certainly doesn't hurt anything.*

THE SIMMERING

> 1 whole country ham, smoked
> cool spring water
> 1 quart Coca-Cola
> 1 teaspoon whole cloves
> 6 allspice berries
> 6 black peppercorns

Scrub the ham and soak it overnight in cool water. Weigh the ham. Cut off the ham hock with a hacksaw if necessary to fit the ham in your pot or roaster. (When choosing a utensil, remember that the ham should be completely covered with liquid.) Put the ham into the pot, then add the Coca-Cola, cloves, allspice, peppercorns, and enough spring water to cover. Bring to a boil and simmer for 25 minutes per pound of ham. Let the ham cool in the liquid.

THE BAKING

> 1 cup Coca-Cola
> boiled ham (above)
> lots of whole cloves
> lots of dark brown sugar

Preheat the oven to 350°F. Select a baking pan or roaster large enough to hold the ham. Put a mixture of the Coca-Cola and 1 cup water into the bottom of the pan. Trim all the skin from the ham, leaving about ½ inch of fat. Stud the ham with cloves spaced ¼ inch apart. Cover the ham (and cloves) with a ½-inch layer of dark brown sugar. Bake for 1 hour. Turn the heat down to 250°F and bake for 30 minutes more. The sugar should caramelize and be dark brown or almost black. Cool and slice thinly.

A British Baked Ham

In Great Britain, a baked ham is traditionally served at Christmas. (Baked ham is also traditional Christmas fare in Sweden and Finland.) The typical British ham is a little sweeter than most European hams. It is sugar-cured, and before sugar was widely available, honey was used in the cure. Molasses is also used. British hams are lightly smoked with oak wood.

1 whole cold-smoked ham
3 cooking apples, peeled and chopped
3 ribs celery with leafy tops, chopped
2 medium onions, chopped
2 medium carrots, chopped
2½ cups hard cider
1½ cups dark molasses
1 cup chopped fresh parsley
12 black peppercorns
dark brown sugar
whole cloves

Put the ham in a large container of cold water. Soak it for at least a day, preferably longer, changing the water several times.

When you are ready to cook, scrub the ham and put it into a container large enough to hold it and enough water to cover it. Add all the ingredients except the cloves and brown sugar. Bring to a boil, reduce the heat, and simmer for 5 hours, adding more water if needed. Turn off the heat and let the ham cool overnight in the liquid.

Preheat the oven to 450°F. Skin the ham, leaving much of the fat on the meat. (The skinning should be done with the aid of a small, sharp knife.) Make crisscross cuts over the fat and insert a clove in the center of each diamond. Sprinkle the brown sugar over the ham and put it into the hot oven for 5 or 10 minutes. Check it a time or two after 5 minutes—and don't let the crust burn.

Leftover Ham Recipes

Ideally, a whole ham, cooked as above, should be put on the table at a festive dinner, such as Thanksgiving or Christmas. What's left over can be eaten in various ways, such as in sandwiches or casseroles and stews. Here are a few of my favorite recipes:

Scotch Mess

My mother made this dish frequently (by request) and called it ham pie. Almost always, it was made with the trimmings taken from around the bone of what was left of a baked country ham. Mother never did measure the ingredients—and I don't, either.

    ham bits and pieces
    dumpling strips cut from rolled pastry
    chicken eggs

    whole milk
    salt and pepper

Hard-boil the eggs, peel them, and cut them into slices. Preheat the oven to 350°F. Trim off some ham, and with it line the bottom of a baking dish or pan about 9 by 13 inches, and about 2 inches deep. (I use a Pyrex dish.) Put a sparse layer of dumpling strips atop the ham, then add a layer of egg slices. Add another layer of ham, dumpling strips, and eggs. Add a layer of ham. Sprinkle with salt and pepper. Add a crisscrossed layer of dumpling strips, then cover it all with the milk.

Bake the pie for about 30 minutes. Then turn on the broiler and lightly brown the top. Serve hot.

Although I don't offer exact measurements for this recipe, the dish benefits from lots of ham and sparse use of dumplings—but, of course, a good deal depends on how much meat you've got and on how many folks you've got to feed. If you don't have a recipe for dumpling strips, try this:

    2 cups all-purpose flour
    2 teaspoons baking powder
    1 teaspoon salt
    ¼ cup vegetable oil
    about ½ cup cold milk

In a medium bowl, mix the flour, baking powder, and salt. Cut in the vegetable oil. Slowly stir in the milk, using just enough to make a rather stiff dough. Put the dough on a well-floured surface and roll it out about ⅛ inch thick. Cut the dough into strips 1 inch wide and use as directed.

Pinto Beans and Ham Bone

While writing this book, I looked at a number of books about the chuck-wagon cooks of the Old West, looking for recipes and ways of cooking salt pork, cured bacon, and so on. I didn't find much information of culinary value, but I did find some good stories. Just the titles of the books were interesting, such as *Shoot Me a Biscuit.* On the trail, of course, the cowboys ate mostly beans, coffee, biscuits, fresh beef, and cured meats. Beans were always available, partly because they traveled well, stored compactly, and didn't spoil.

In one of the books, a cowboy fresh off the trail went into a swanky restaurant in St. Louis or somewhere. He couldn't read the menu, so, after some hemming and hawing, he asked the waiter to identify all the entries that contained beans. The waiter spoke, but the words didn't help.

"Put your finger on them that contains beans," the cowboy said. Somewhat impatiently, the waiter put his finger on a single entry, knowing that he wasn't going to get much of an order or tip from anyone looking for beans. "That's the only one?" the cowboy asked. The waiter assured him that it was the only bean dish available at the establishment. "Well, you hold this one," the cowboy said, putting his left index finger on the entry and sweeping his right hand over the entire menu, "and bring me an order of all the rest of this stuff."

I don't claim that my pinto bean recipe would have sold that cowboy on the idea of eating another spoonful of beans, but I do recommend it to anyone who has been eating pinto beans that have been soaked all night in water and boiled without a good ham bone.

1 pound dried pinto beans
1 ham bone, with some meat on it
½ teaspoon hot red pepper flakes
salt, if needed
1 sprig epazote, if available

Put the ham bone into a large pot or Dutch oven. Pour in the beans and cover well with water. Add the red pepper flakes and epazote. Bring to a boil, reduce the heat to a simmer, cover tightly, and cook for 6 or 7 hours, adding more water from time to time if needed. Season with salt, if needed.

Serve hot with sliced Vidalia or Texas onions and corn pone made from stone-ground meal. If the ham bone has lots of meat left on it (as well it should), I can make a whole meal of this dish. Leftover beans can be mashed, shaped into patties, and fried on a griddle.

Ham and Swiss Cheese on Rye

At our house, most leftover ham is used in sandwiches. Usually, these are made as needed or as wanted, and require nothing more than white sandwich bread and mayonnaise. For a light meal, my wife and I make a sandwich of Swiss cheese, rye bread, Dijon or Creole mustard, lots and lots of thinly sliced ham, and maybe a little lettuce. The thinner the ham, the better your sandwich will be. Cut the sandwiches in half and serve with thick potato chips and huge dill pickles.

After the first edition of this work came out, a critic asked why anyone would spend good money on a book that contained instructions for making a ham sandwich.

Well . . . after much hemming and hawing, turning and tossing, I have left the sandwich suggestions in the revised edition of this modest work, simply because some people (never mind who) really do need some help, not only with a deli-type rye-bread sandwich, but even with an ordinary white-bread ham sandwich.

Here's my advice. Use very fresh white sandwich bread. Smear two slices quite liberally with a good mayonnaise. Then sandwich a dozen or so pieces of thinly sliced ham. No lettuce.

It's that simple. Try it. Cut the sandwich in half diagonally. Pick up a half in your right hand and look at it. Go ahead. Take a bite right out of the middle. This one bite will be worth more than the small price of this little book.

COLD-SMOKED BACON

The procedure in Part One for making salt pork can be extended for cold-smoked bacon. After the cure, wash the belly slab, dry it, hang it in a smokehouse, and cold-smoke it for at least a full day, or up to a week. Some people like honey and pepper on their bacon. I think it's best to add these ingredients after the smoking is completed.

Sugar Cure for Bacon

If you want a good sugar cure especially for bacon, try this old British formula, adapted from The Country Kitchen:

approximately 12 pounds bacon
1 pound salt

COLD-SMOKING & SALT-CURING MEAT, FISH, & GAME

  1 pound brown sugar
  1¾ pints wine vinegar
  1 ounce saltpeter
  2 tablespoons black peppercorns

Mix the salt, brown sugar, peppercorns, and saltpeter, then divide
the mixture in half. Rub half the cure into the meat, covering
all surfaces nicely. Leave the meat on a wooden shelf or table in
a cool place, turning daily. After 5 days, mix the rest of the cure
ingredients with the wine vinegar. Rub the bacon with part of
this mixture. Turn daily, rubbing again with part of the liquid, for
5 days or so, or until the cure has been used up.

Punch holes in 2 corners of the bacon, then insert cotton
string or hickory withe. Hang the bacon in a smokehouse or
cold-smoker for at least 24 hours, preferably for a week or so.
This bacon will keep for several weeks at room temperature, and
longer in the refrigerator. It can also be frozen.

Pioneer Bacon

Here's an old recipe adapted from Cooking in Wyoming, *which
contains a chapter or two on pioneer cooking.*

  80 pounds side meat for bacon
  plenty of salt
  ½ cup brown sugar
  2 ounces saltpeter
  black pepper
  red pepper
  starch

154

Pulverize the saltpeter, then mix it with the brown sugar and 4 cups of salt. Dampen the mix slightly with a little water and rub it all over the meat. Lay the meat skin side down on boards for 9 days. Then rub thoroughly with salt; use all the salt that will adhere to the meat. Let lie for 3 days.

Hang the side meat in the smokehouse and cold-smoke it for 3 days. Sprinkle each piece with black pepper. Have ready a cloth sack for each slab of bacon. Mix a solution of starch and water, adding a little red pepper to prevent mold and keep the flies away. Wet the sacks with the starch solution, making them stiff. Put each piece of meat in a cloth sack, tie the end, and hang in a cool place. (In warm climates, it's best to refrigerate the bacon if it is to be kept very long. Bacon, being fatty, tends to go rancid if kept too long after curing and smoking.)

COLD-SMOKED SAUSAGE

Be warned that smoked sausage can be dangerous, leading even to deadly botulism. The problem, as I see it, is that (1) the meat is not normally given a proper salt cure and (2) the meat-grinding process tends to spread any bacteria that might be present on the surface. With pork sausage, the inherent fat may work to your advantage because lard is a known preservative. In other words, fatty sausage may be safer than lean sausage. Also, the hog casing somewhat protects the surface of the sausage. But there may be other opinions on this matter. Since whole books have been written on this subject, anyone serious about sausages should make a trip to a good public library. These books should be read carefully—and questioned at every turn. My best advice on smoked sausage is that it be cooked thoroughly and refrigerated or frozen as soon as the smoking period is over. If you want to keep the sausage without refrigeration, it's best to cook it, layer the cooked links in a clean crock or other suitable container, and cover it with lard. This is an old-timey method of preserving sausage.

Smoked Sausage

I don't remember how my father preserved our sausage, but I do remember that (1) it wasn't left in the smokehouse for very long and that (2) it was very, very good. A local character of my acquaintance was a sort of overseer for a large farm run by a Ford dealer, who was said (at one time) to have sold more Ford tractors than any other man in the world. The dealer liked to take this character on long trips, and they always sampled the vittles in the local restaurants, swanky or otherwise. One day I heard him say that he had eaten in the Forum of the 12 Caesars in New York City, in Los Nova Dados in Tampa, on Fisherman's Wharf in San Francisco, and at Antoine's in New Orleans. "But the best food I've ever eaten," he said, licking his lips, "was corn syrup, biscuits, and sausage."

I agree that good sausage is hard to beat for brute flavor. Here's my recipe, made with rather lean meat. Traditionally, pork sausage is made from all the trimmings, but some of the better sausage is made with the whole hog and is called whole-hog sausage. Some farmers of my acquaintance boast of putting a whole ham or two in with the sausage.

    10 pounds pork with about 20 percent fat
    ¼ cup finely ground sea salt
    2 tablespoons black pepper

2½ tablespoons dried sage
1 to 2 tablespoons red pepper flakes

Mix all the seasonings. Chill the pork, cut it into small chunks, and sprinkle it with the seasonings. Run the pork through a sterilized sausage grinder, using a fine wheel. Stuff the ground meat into pork casings, hang the links in a smokehouse, and cold-smoke for 2 or 3 days. Cook the sausage thoroughly before eating it.

COLD-SMOKED BIRDS

As a rule, the larger the bird, the more time it takes to cure and smoke it properly. In all cases, cold-smoked birds should be cooked before they are eaten. In most cases, it's best to pluck the birds instead of skinning them.

Pheasant and Chickens

Although the meat is on the dry side, pheasant can be salt-cured and cold-smoked successfully. Start by plucking the birds and removing the heads, feet, and innards. Mix a brine in the proportions of 2 cups salt, 1 cup sugar, and ¼ cup crushed allspice berries per gallon of water. Put the birds into a crock or other suitable nonmetallic container. Pour the brine over the birds, covering them. Then weight the birds with a wooden block or some such object so that they are completely submerged. Leave the birds in the brine for 10 days.

Remove the birds, wash them with clean water, and hang them to dry for several hours. Then cold-smoke the birds for at least 2 hours. The longer they smoke (up to a point), the better the color and flavor will be—and the dryer the flesh. Birds to be baked can be smoked for a couple of hours, barded with bacon, wrapped in aluminum foil, and baked until done. Birds to be used in stews and soups, or cooked by any wet method, can be smoked longer for more flavor.

Chickens can be smoked by this method, but I personally don't want to start out with supermarket birds or birds that have been raised in compartments, cleaned in a mass operation, and run through a salmonella bath. My concern is freshness and salmonella, plus the fact that most commercial birds are too fat. I eat the birds purchased at the supermarket, but it's best to cook them as soon as possible. I don't even want them in my refrigerator these days. There are surely other feelings about this matter; some people can't bear the thought of eating game birds or barnyard chickens that aren't inspected by the U.S. Department of Agriculture.

Quail and Small Birds

Quail are very good when cold-smoked, and they are easy to prepare. Simply soak them in a brine (the one for pheasant will do) for a couple of hours, dry them, cold-smoke them for 3 or 4 hours, and cook them by a recipe of your choice. I especially like quail baked with mushroom soup and rice. I am fond of wild quail, but I also like the pen-raised quail for smoking. If you don't grow your own quail, look for someone in your area who does. I have also been successful with frozen quail from the supermarket.

Other small game birds, such as doves and snipe, can be cold-smoked before cooking. (Personally, however, I prefer to smoke game birds with white meat.) Grouse and Cornish game hens can be smoked, but you'll need to extend the curing and smoking times.

Turkey

I much prefer wild birds (which, contrary to much opinion, are not as dry, at least before cooking, as domestic birds) or birds that have been raised on the ground (with plenty of room to scratch) instead of in a compartment. A happy bird makes better eating.

More and more these days people are raising their own turkeys, so it shouldn't be too difficult to find a local grower who will sell you a live bird or two. After you chop off its head, draw and pluck it as soon as possible. Wash the bird quickly with salted water, then weigh it and record the figure. Place the bird in a crock or other suitable container and cover it with brine. (The brine formula used for pheasant will do.) Weight the bird with a block of wood or other suitable nonmetallic object so that it stays completely submerged. If the bird weighs less than 10 pounds dressed, leave it in the brine for 1 day per pound; if it weighs between 10 and 15 pounds, 1¼ days per pound; over 15 pounds, 1½ days per pound. It is important also that the bird be cured in a cool place, preferably at 38°F. Every 7 days or so, remove the bird (or birds) and stir the brine about, and put the bird back into the brine.

After completing the cure, wash the bird inside and out in fresh water, then hang it to dry. I hang it with a cord tied to each foot, but some people tie a cord around each wing and suspend the bird with the breast down; still others prefer to put the bird in a net bag for hanging. After the bird dries hang it in the smokehouse and cold-smoke it for at least 2 days, until the skin is a mahogany color. The longer it is cold-smoked, the stronger the flavor—and the dryer the meat it will be. After being cold-smoked, the bird should be aged in a cool place for a week or so.

Of course, a turkey smoked by this method should be thoroughly cooked before it is eaten. For baking, preheat the oven to 350°F, wrap the bird with strips of bacon, insert a meat thermometer into the thickest part of the thigh (without touching bone) or the breast, and bake the bird until the internal temperature reaches 180°F. If I know that I have a good bird, I'll turn off the oven heat when the thermometer reads 160°F, then let the bird coast for an hour or so as the oven cools down. Turkey is best, at least culinarily, when it is cooked to only 145°F or 150°F, but the U.S. Department of Agriculture recommends 180°F, a figure that I feel obligated to report here.

It is possible, of course, to cold-smoke a turkey for a couple of hours and then put it in the oven to cook. The results may be quite tasty, but the deep, rich smoke flavor may not be present in the meat except on the surface.

Wild Ducks and Geese

These days, smoked wild ducks are usually hot-smoked during the cooking process. It is possible, however, to cold-smoke ducks, getting a more intense flavor. Here's a report by Dr. Andrew Longley from Cundy's Harbor, Maine, as published in *The Maine Way*:

Prepare ducks by plucking and dressing as one normally would for baking. Soak the ducks overnight in salted water.

Smoke in smokehouse for 36 to 48 hours. (For optimum results, smokehouse should be maintained at 80 degrees.) After smoking is completed, ducks are baked as one would normally prepare ducks. Whistlers, buffleheads, and black ducks have been smoked with good results.

I recommend that the ducks be drawn, plucked, and salted as soon as possible after the kill. For the brine soak, I recommend at least 2 cups of salt per gallon of water.

The above method also works for wild geese. (It's best to proceed with a young bird unless you are going to cook it by long stewing.) Soak the bird in brine for 2 days, smoke it for 3 or 4 days, and then cook it with your favorite recipe. I am especially fond of a slowly cooked gumbo made with smoked goose.

COLD-SMOKED SMALL GAME

Rabbits and other small game can be cured and smoked quite successfully. Opossum is especially good for smoking. For animals of cottontail size, wash and soak them in a light brine for several hours, put them in a pickle for 5 or 6 hours, then cold-smoke them for a day or so. Whole hares and larger domestic rabbits should be pickled for 10 hours or so, or dressed and pickled the same as cottontails.

COLD-SMOKED VENISON AND LARGE GAME

Most game tends to be on the dry side as compared to pork and feedlot cattle. Even so, it can be pickled and cold-smoked. In Russia, for example, bear hams are salt-cured and smoked exactly like pork. The hams, shoulders, and saddle of deer smoke nicely, if you first use a sweetened pickle in the proportions of 1 pound salt, ½ pound brown sugar, 2 tablespoons sodium nitrite (optional), and ½ cup juniper berries per gallon of water. If you have a whole deer and a large pickling container, put the hams on the bottom, the saddle in the middle, and the shoulders on top. (Some people recommend that the deer be aged for a week or so before starting the pickle, but I don't think this is necessary if you have a good deer that was cleanly killed and promptly field-dressed.) You can also add the ribs and perhaps the tenderloin along with the ribs, but I like to start eating on these right away. Be sure to keep the meat completely submerged in the pickle. After 3 days, remove the top layer or two, dry the meat, and start cold-smoking. Turn the other meats and stir the pickle. After another 3 days, remove the shoulders, dry them, and start cold-smoking. After another 3 days or longer, remove the hams, dry them, and start cold-smoking. The shoulders should be smoked for 1 week, the saddle for 1½ weeks, and the hams for 2 weeks or thereabouts. During cold-smoking, it's best to rub the venison from time to time with bacon drippings.

Note that the curing and smoking times do not have to be exact if the meat is cooked or frozen shortly after cold-smoking. Note also that large game such as elk and moose can be cured and smoked, but I think it better to separate the rear leg into several roasts instead of trying to cure the whole thing.

COLD-SMOKED MUTTON HAM

At one time, this dish was quite popular in northern Europe, and it is still eaten in some countries, especially Norway. This recipe has been adapted from *The Country Kitchen,* a delightfully

British book. Mutton has a stronger flavor than lamb, and is not often marketed in the United States. This is a pity, but lamb can also be used. It's best, of course, to proceed with very fresh meat. Anyone who has a few acres should look into raising lamb and mutton for food. They are relatively easy to butcher as compared to beef, and they don't have to be scalded and scraped like hogs.

Smoked Mutton Ham

> 1 leg lamb or mutton
> ½ pound sea salt
> ½ cup dark brown sugar
> 1 tablespoon crushed black peppercorns
> 1 tablespoon crushed coriander seeds
> 1 tablespoon crushed allspice
> 2 teaspoons saltpeter
> wine
> chopped celery
> chopped onions
> chopped garlic

Thoroughly mix the salt, brown sugar, saltpeter, coriander, all-spice, and peppercorns. Rub the fresh leg of lamb all over with the salt mixture, pushing it in around the exposed bone. Place the leg of lamb on a nonmetallic surface and put it in a cool place for 10 to 14 days. Turn the leg daily and rub the top side with the liquid that accumulates on the bottom. After 10 to 14 days, dry the ham and cold-smoke it for a day or two.

When you are ready to cook, place the ham in a large container and cover it with a stock made of 4 parts water to 1 part wine, along with some chopped celery, onions, and garlic. Bring the liquid to a boil, reduce the heat, and simmer for 4 hours, adding more liquid from time to time if needed. Let the ham cool in the liquid. Then remove the ham, cover it with a clean cloth, lay a wooden board on top, and weight it overnight with a flat iron or two.

COLD-SMOKED TONGUE

Cured and smoked beef tongue is one of my favorite delicacies. It's best to cure several of these at a time, and they should be purchased fresh at a meat processing plant. Wash the tongues in salted water, then pack them loosely in a nonmetallic container. Cover them with brine (at least 2 cups of salt per gallon of water) and weight them down with a plate, board, or some nonmetallic object. Leave the beef tongues in the brine for 3 or 4 weeks, repacking and stirring the brine every week or so. Dry the tongues and cold-smoke them for a day or longer.

To cook, cover the smoked tongue with water, then add a few juniper berries, peppercorns, and celery tops; boil the tongue for 6 or 7 hours, or longer, until the meat is very tender. Let the tongue cool, peel it, and slice it crosswise.

I can make a meal of smoked tongue, but it is more often served as an appetizer. It is good with stone-ground brown mustard and thin slices of party bread. Sliced tongue also makes a tasty sandwich. If your guests tend to be on the squeamish side, always slice the tongue crosswise and arrange it in a nonsuggestive manner. Otherwise, it might not be an appetizer.

Tongues from sheep, deer, and other mammals can also be cured and smoked to advantage. Tongues from elk and moose should be about the same size as beef tongues, and should be cured and smoked accordingly. For smaller tongues from deer and sheep, reduce the curing and smoking times by about 70 percent. Whale tongue—an old Basque delicacy—should be cured and smoked considerably longer.

African Safari Biltong

In South Africa, chunks of beef are moistened with vinegar, rubbed with salt, pepper, and coriander seed, hung in the wind for a few days, and finally hung in the chimney to smoke.

In the bush, game is dried with the same seasonings, but the meat is cut into thin strips and hung to dry like American jerky. Zebra, it is said, makes the best possible bush biltong.

METRIC CONVERSION TABLES

Metric U.S. Approximate Equivalents

Liquid Ingredients

Metric	U.S. Measures	Metric	U.S. Measures
1.23 ML	¼ TSP.	29.57 ML	2 TBSP.
2.36 ML	½ TSP.	44.36 ML	3 TBSP.
3.70 ML	¾ TSP.	59.15 ML	¼ CUP
4.93 ML	1 TSP.	118.30 ML	½ CUP
6.16 ML	1¼ TSP.	236.59 ML	1 CUP
7.39 ML	1½ TSP.	473.18 ML	2 CUPS OR 1 PT.
8.63 ML	1¾ TSP.	709.77 ML	3 CUPS
9.86 ML	2 TSP.	946.36 ML	4 CUPS OR 1 QT.
14.79 ML	1 TBSP.	3.79 L	4 QTS. OR 1 GAL.

Dry Ingredients

Metric	U.S. Measures	Metric	U.S. Measures
2 (1.8) G	1/16 OZ.	80 G	2⅘ OZ.
3½ (3.5) G	⅛ OZ.	85 (84.9) G	3 OZ.
7 (7.1) G	¼ OZ.	100 G	3½ OZ.
15 (14.2) G	½ OZ.	115 (113.2) G	4 OZ.
21 (21.3) G	¾ OZ.	125 G	4½ OZ.
25 G	⅞ OZ.	150 G	5¼ OZ.
30 (28.3) G	1 OZ.	250 G	8⅞ OZ.
50 G	1¾ OZ.	454 G 1 LB.	16 OZ.
60 (56.6) G	2 OZ.	500 G 1 LIVRE	17⅗ OZ.

GLOSSARY

Not all of the terms below are used elsewhere in this book; however, they may be helpful to readers coming here from other books or who might otherwise expect to see the terms. A few terms, such as "salt equalization," may not be used in other books on the subject. Most of the terms are discussed in more detail in the text, to which the index is the best guide.

Age To hang meat in a dry place after curing. If the meat has been properly cured, the temperature at which the meat is aged is not critical. Proper aging gives a distinct flavor to country hams and other meats.

Ascorbic Acid A form of vitamin C that is used in canning to preserve the color of fruits and vegetables. It is sometimes added to a meat cure to help retain the color of the meat.

Botulism A deadly food poisoning that is usually brought about by improper canning of meat, fish, and vegetables. In order to multiply to dangerous levels, the bacteria, which produce a toxin, require moisture and a lack of oxygen. Botulism is not often a problem in cured meats, except when canned. It may, however, occur in some cased sausages (made with meat that hasn't been salt-cured) and possibly in other moist meats.

Brine Salty water used for curing meats and fish. Ordinary salt is the major element, but other ingredients may be added for flavor or color. An egg will float in a strong brine, and this fact is often used to measure the strength of a brine. A salinometer can also be used. Note that the salt in a brine draws moisture from meats, so that the brine can become diluted during the curing process unless more salt is added. Brines used to cure meats should be stirred from time to time, which may require the meat to be removed and repacked; this process is called overhauling. For culinary purposes, the water used in the brine should be of good flavor.

Cold-Smoking Smoking meats or fish at a low temperature for a long period of time. The temperature should be below 100°F, and preferably between 70°F and 80°F. Bacteria multiply rapidly at temperatures between 80°F and 140°F.

Cure A wet or dry mixture containing salt and possibly other ingredients. Salt is the critical element. The word *cure* is sometimes used to denote sodium nitrite, sodium nitrate, potassium nitrite, or potassium nitrate. As a verb, *cure* means simply to subject to a cure for a period of time. Aging has a separate meaning. Cured meats can be aged, in which case temperature is not as important as in curing.

Dry Cure A meat cure consisting of dry ingredients, mostly salt. As a verb, *dry-cure* means to cure with dry ingredients.

Honey A natural preservative, honey is sometimes used instead of sugar in a sugar cure. It is also applied to the surface of some meats, especially bacon.

Hot-Smoking Flavoring meats or other foods with the aid of smoke during the cooking process. Hot-smoked foods can be eaten immediately after the simultaneous cooking and smoking is over, but these foods are not cured or preserved.

Hygrometer A device used to measure the humidity inside a meat-curing chamber. It can be important for commercial applications, where exactness is important to duplicate results from one batch to another. As a rule, ham curing and salt equalization should be accomplished at a relative humidity of 75 to 90 percent; ham aging, on the other hand, is best at 50 to 60 percent.

Injection Pump A syringe designed for injecting liquid cures into large pieces of meat.

Kosher Salt A coarse salt that can be used for pickling.

Nitric Oxide When sodium nitrite is applied to meat, it breaks down into a chemical called nitric oxide, which is the stuff that really does the work.

Overhauling Removing meat from a brine, stirring the brine, and putting the meat back into it. Sometimes more salt is added to the brine at this point.

Pellicle The thin, shiny layer that forms on the surface of fish that has been soaked in brine and then dried for several hours in a breeze. A good pellicle improves the appearance of the final product, and may aid in preserving the fish.

Pickling Salt A salt that contains no iodine. It is sold commercially for use in pickling, canning, or meat curing or brining.

Potassium Nitrate Saltpeter. In the past, saltpeter has been widely used in meat cures and in gunpowder. Its use in commercial meats has been curtailed considerably.

Prague Powder A trade name for a curing mixture of salt and other chemicals. Prague Powder 1 contains salt and odium nitrite. Prague Powder 2 contains salt, sodium nitrite, and sodium nitrate.

Salinometer A device for measuring the salinity of a liquid. It can be used to determine the salt content of a brine used in meat curing.

Salmonella Bacteria that can cause food poisoning. To reach dangerous levels in meats, it requires moisture and the right temperature window. Salmonella is especially a problem with modern mass-produced poultry, although the bacteria can also multiply in other meats and fish, as well as in eggs. It is not a problem with meats that have been properly cured and fully cooked.

Salt Sodium chloride. The term "curing salt" is sometimes applied to sodium nitrite and other chemicals. There are several kinds of salt, depending on how it is mined or processed; most salts are not pure sodium chloride. Any natural salt can be used for curing meats. Salt also comes in various grains, from fine to large chunks commonly known as ice cream salt. Any of the forms can be used in a brine, provided it is well mixed, but salts for dry cures work best in a fine grain.

It has been noted that man did not use sodium nitrite and other minerals in his cured meats until quite recently. This is not the whole story. Until recently, these minerals were not removed from natural sea salt before it was used for curing meats. Modern man is, in short, taking sodium nitrite and potassium nitrate and other minerals out of salt—and then putting them back into the salt used to cure meats. In any case, my favorite salt for curing and table use is unrefined sea salt, simply because it has good flavor. Sea salt is, however, too expensive these days for large-scale meat curing.

Salt Equalization When a piece of meat is salted, at least two things happen: Moisture is drawn out of the meat, and salt penetrates into it. At first, only the outside of the meat will be salty. In time, the salt will penetrate deeper and deeper into the meat, tending to equalize the salt content. Salt equalization in a large ham may require 2 months, which includes a curing period of about 40 days and a salt equalization period of about 20 days. Salt equalization is a very important step in curing country hams.

Saltpeter Potassium nitrate. It is used in meat curing and in explosives. It is poisonous and should be used carefully in small amounts, if at all. Its use in commercial meats has been curtailed considerably.

Sodium Nitrate A sodium nitrogen salt that is widely used in meat cures as well as in fertilizers and explosives. Because it is toxic, sodium nitrate should be used carefully, in small amounts, and should be mixed thoroughly with salt. Both federal and state agencies limit the use of sodium nitrate in commercial meats.

Sodium Nitrite A salt of nitrous acid that is widely used in meat cures. Because it is toxic, sodium nitrite should be used carefully, in small amounts, and should be mixed thoroughly with salt. Both federal and state agencies limit the use of sodium nitrate in commercial meats.

Sugar Cure A mixture of salt and sugar for meat or fish, although salt is the essential ingredient. Up to a point, sugar gives the ham a better flavor and a good color, but too much sugar makes the meat slimy.

Trichina A microscopic worm found in the flesh of hogs and bears. When ingested, it can cause serious illness. Fully cooking the meat kills trichina.

INDEX

ABOUT THE AUTHOR

A. D. Livingston claims to have hopscotched through life. Navy at seventeen. Mechanical engineering at Auburn. Atomic bombs at Oak Ridge. Creative writing at University of Alabama. Missiles and rockets at Huntsville. Published a novel and played a little poker. Travel editor at *Southern Living* magazine. Freelance writing and outdoor photography. Word man for fishing rods and bait-casting reels with Lew Childre, the genius of modern fishing tackle. Bought the family farm. Lost the back forty publishing *Bass Fishing News*. Lost the rest of the farm manufacturing fishing lures. Back to freelancing. Published twenty-something books. For the past sixteen years—the sweetest of all, he claims—he has been the food columnist for *Gray's Sporting Journal*. What in his previous work experience qualifies him for this position? Nothing whatsoever. He hates to work, but all his life he has loved to hunt and fish and to cook and eat the bounty. And he loves to write about it his way.